ENDORSEME

Emotionally Fit is not a self-help manual nor a catalog of easy advice. It is an invitation to real, deep, and transformative training. A process that begins where many never dare to look: inward.

— César De León, PhD. Director of the Family Ministries Department, North American Division

In a world where many men are striving but not fully understanding themselves, Emotionally Fit offers clarity, direction, and transformation. This powerful work uncovers the deep needs of every man—respect, purpose, belonging, achievement, and transcendence—and reveals how, when misunderstood, they can lead to destructive patterns, but when properly channeled, produce strength, discipline, and godly character.

Blending insights from neuroscience, psychology, sociology, and theology, this book equips men to understand how their minds work, master their emotions, and reshape their habits. More than a book, it is an eight-week journey designed to help men take control of their inner world and step into their highest potential.

**— Alanzo Smith, PhD. LMHC, LMFT.
President: Greater NY Conference**

I enjoyed reading the book Emotionally Fit. Dr. Samuel Peguero, with clarity and academic depth, presents a practical path for men—whether married or not—to reach their highest potential.

— Jorge Mayer – Adjunct Professor at Andrews University

Emotionally Fit is a deeply necessary and timely work that courageously addresses one of the most urgent and least discussed topics in contemporary society: men's emotional health. The author demonstrates an exceptional balance between academic rigor and practical applicability, integrating biological, psychological, sociological, and theological perspectives in a coherent and accessible way. The book also stands out for its honesty in addressing taboo topics such as pornography, sexual addiction, uncontrolled anger, and male emotional incapacity. The author does not minimize these problems nor offer magical solutions, but rather presents a realistic path of growth based on neuroscience, psychology, and spiritual principles.

—Jobnel Tineo – Director of Men's Ministries, Texas Conference

I highly recommend reading Emotionally Fit by Dr. Samuel Peguero. In a culture where many men face their emotional struggles in silence, this work provides biblical guidance and practical tools for holistic development. It is a call to strengthen character, emotional health, and spiritual leadership. It will be a great blessing for those who long to live with purpose and maturity.

—Mauricio Buitrago, DMin. Ministerial Association, North Colombian Union

Reading The Emotionally Fit Man has been a revealing experience. Every man who has begun his process of emotional growth will find here clear guidance and concrete exercises to develop a more conscious and balanced life. Likewise, every woman who desires to better understand and accompany the male emotional world will find in this work valuable keys to walk more wisely alongside him.

— Sadis Valencio – Psychotherapist, writer

In this book, nothing is left to the imagination. The author explains the importance of making decisions at the right moment and in a conscious manner; he also highlights how an 'amygdala hijack' could cause damage in the short, medium, and long term.
He emphasizes how emotional and spiritual reeducation, balance, emotional regulation, neuroplasticity, among other aspects, make it possible to retrain the brain and obtain tangible results, with noticeable changes in behavior.

— Karina De la Rosa de Díaz – Family Counselor

EMOTIONALLY FIT

EMOTIONALLY FIT

How to Train Your Brain for Purposeful Masculinity

Samuel Peguero

Emotionally Fit: How to Train Your Brain for Purposeful Masculinity
Title of this book in Spanish:
Emocionalmente Fit: Como Entrenar Tu Cerebro Para Una Masculinidad Con Proposito

Author: Samuel Peguero

ISBN: 978-1-972915-38-7

Publisher: Stilo Media
Publication Date: July 2026

Layout: Had Graphic Inc.
Cover Design: Lisbeth Peguero Pena & Héctor Delgado
Cover Image: Juan Joaquín
Proofreader: J. Juana Ramirez

Contact Information
Email: samuelpeguero1@icloud.com
YouTube: Dr. Samuel Peguero
Instagram: @pegueroam
Facebook: Samuel Peguero
www.hopenestfamily.com

Available in Spanish, French, and Portuguese in both eBook and print formats.

TABLE OF CONTENTS

Endorsements / i
Introduction / xi
Foreword / xiii
In Gratitude To: / xv
Initial Self-Assessment / xvii

01

LEND ME YOUR LENS 01

02

IDENTITY AND PURPOSE 17

03

CONTROLLING THE UNCON-TROLLABLE 33

04

TO SPEAK OR NOT TO SPEAK 51

05

THE MAN AND HIS SEXUALITY 65

06

MAN AND MONEY: LET'S TALK SERIOUSLY... 85

07

THE APPARENT FEMALE MYSTERY 107

08

INTEGRALLY FIT 133

Emotionally Fit 157
Final Assessment 158
Other Works by the Author 165
About the Author 175

INTRODUCTION

Being a man in today's world implies more than fulfilling external responsibilities; it demands an inner strength capable of sustaining decisions, relationships, and purpose. Many men strive to develop their bodies, their careers, and their ability to provide, but they have rarely been trained to understand and govern their emotional world. *Emotionally Fit* was born precisely to fill that gap. It is not a call to weakness, but to true strength—the kind that arises when a man understands how his mind works, becomes aware of his triggers, and learns to direct his emotions instead of being directed by them.

From a psychological perspective, the man who does not understand his emotional life tends to react instead of respond. Anger, frustration, anxiety, or defensive silence end up affecting his marriage, his fatherhood, his leadership, and his inner peace. In contrast, the emotionally trained man develops greater self-control, makes more strategic decisions, improves his communication, and strengthens his self-esteem. When his internal world is in order, his external world will align. Reading this material means investing in your stability, your moral authority, and the quality of your relationships.

Man has deep needs such as respect, purpose, belonging, achievement, and a sense of transcendence. When these needs are not understood or properly processed, they can become distorted and turn into impulsive or self-destructive behaviors. But when they are understood and properly channeled, they transform into creative energy, balanced leadership, and firm character. Being emotionally "fit" is not optional; it is essential for anyone who desires to live with integrity, influence, and fulfillment.

This work integrates principles from biology, neuroscience, sociology, psychology, and theology to offer a scientific and practical explanation of the male brain. Here you will discover how the amygdala and the prefrontal cortex operate, how habits are formed, how social learning shapes your identity, and how spiritual principles strengthen self-control. You will learn to work on your weaknesses, enhance your strengths, and reprogram those areas that have limited your growth.

Emotionally Fit is not just a book; it is an eight-week training journey for men, including the first and last chapters as opening and closing. It is an invitation to understand your design, take control of your mind, and become the best version of yourself. The man who governs his inner world is in a position to govern his destiny. In addition, when a man is emotionally strong, his family, his community, and his legacy will be as well. Are you ready? Then let's start training right away!

Samuel Peguero
Author

FOREWORD

There is a question that, sooner or later, every man asks himself in private. Not out loud, not in front of others, but in those quiet moments when the noise of the world cannot reach him: *Is this all that I am?* It is not a question of weakness. It is the bravest question a man can ask himself. This book was born to accompany him in finding an honest answer.

Emotionally Fit is not a self-help manual or a catalog of easy advice. It is an invitation to a real, deep, and transformative training. A process that begins where many never dare to look: inward. Because the man who does not know himself will hardly know those he loves, and the man who does not govern his inner world will end up being governed by it.

Throughout these pages, we will explore man from multiple perspectives: the biology that distinguishes him, the sociology that shapes him, the psychology that reveals him, and the theology that defines him. We will discover how his brain works under pressure, how his emotions are formed, why he reacts the way he does, and most importantly, how he can change. Because change is possible. Science confirms it, and faith guarantees it.

Today's man faces battles his grandfathers never imagined. He struggles with chronic exhaustion and digital loneliness. He carries expectations no one taught him how to fulfill. He faces identity crises in silence, because he learned from a young age that a man "does not show weakness." And in that silence, many have lost marriages, families, health, and purpose—not for lack of intention, but for lack of tools.

This book is that set of tools you need.

But it goes beyond technique. Because at the end of the journey, we will discover that the true model of masculinity is not dictated by culture nor defined by social curriculum. It is embodied in Christ: firm and tender at the same time, powerful and compassionate, a leader who becomes great by serving. That is the direction of every chapter, every exercise, every reflection in this work.

As psychologist Abraham Maslow points out, human beings live in a constant search for: security, belonging, esteem, and purpose. Man is no exception. And when these needs are understood and properly channeled, they cease to be forces that consume him and become energy that drives him.

Being integrally fit is not a destination reached all at once. It is a daily practice, a decision renewed every morning, a journey worth taking. It is training the brain to respond instead of react, to serve instead of dominate, to connect instead of withdraw. When a man is emotionally healthy, his family perceives it, his community notices it, and his legacy—silent but firm—reaches generations.

César De León, PhD, LMFT
Director of Men's Ministries
North American Division of Seventh-day Adventists

In Gratitude To:

I have reserved this page to express my deepest gratitude to a group of people who have been key pieces in the development of this project.

As I look at the final product and the process required to get here, I cannot help but acknowledge the hand of a powerful God who was stronger than the circumstances, who worked with unique wisdom and guided me with love to this point.

Likewise, I recall that one afternoon in Manhattan, NY, while heading to an appointment with a colleague, he handed me the phone. On the line was Jobnel Tineo, Coordinator of Men's Ministries for the Hispanic area in Texas Conference, who challenged me to develop material addressing men's emotional life. Although the idea was not entirely clear at the time and I had many responsibilities, I accepted the challenge. As I progressed, the project grew on me, and I came to love it wholeheartedly.

After completing the manuscript, there was a select team of professionals, including physicians, sociologists, psychologists, counselors, and family therapists, as well as graduates in the field of theology, who carefully reviewed the material in whole or in part. Among them were Doctors: César De León, Alanzo Smith, Ronald Rojas, Efraín Duany, Mauricio Buitrago, Jorge Mayer, Yohanna Morillo, and Manuel Rosario. Likewise, other distinguished professionals such as: Jobnel Tineo, Sadis Valencio, Luz María Reyna, and Karina de la Rosa. Their enriching advice was highly appreciated.

The cover was produced by my daughter Lisbeth, but there was also a wonderful team contributing images and ideas. Juan Joaquín and Héctor Delgado—the effort of the three is highly appreciated.

On this page, I want to highlight the great contribution of my wife and my two daughters. As a man, they have allowed me to play the role of husband and father. They have changed my life more than you can imagine. It has not been an easy task. In fact, I am still a product in the process of being completed, but they have played a special role in the construction of the man I am. Without that foundation, it would have been very difficult for me to embark on this project.

I would lack space to express gratitude to so many men, including my father, who in one way or another provided me with a model of masculinity that has influenced who I am today.

I still have a special space in my heart that I want to pour into this page of gratitude, and it is reserved for you. There, I want you to write your name _________. I know that the intention that moves you to acquire, read, and share this book with your friends is born from the noble desire to help build a masculinity with purpose and with a sense of eternity. Thank you! Thank you very much!

The Author

____

Initial Assessment

Emotionally Fit

Personal Diagnosis of Male Emotional Development

Name: ____

Date: ____

Age: ____

Marital Status: ______________________________________

Instructions

Read each statement and mark the number that best describes your current reality.

Score	Meaning
1	Never
2	Rarely
3	Sometimes
4	Frequently
5	Almost Always

Answer honestly. This assessment is not intended to condemn you, but to help you clearly recognize where you are, what you need, and where you need to grow.

I. Self-Perception and Male Identity

Item	1	2	3	4	5
I have a clear understanding of who I am and the kind of man I want to become.					
My personal worth does not depend exclusively on my success, money, achievements, or external approval.					
I recognize my strengths without becoming prideful.					
I recognize my weaknesses without denying them or making excuses.					
I am able to admit when I need help.					
My identity is guided by principles and not only by impulses or momentary emotions.					

Subtotal: ______ / 30

II. Emotional Management

Item	1	2	3	4	5
I can clearly identify what I feel.					
I express my emotions without aggression or avoidance.					
I manage anger without hurting others.					
I am able to regulate myself when I feel frustrated, rejected, or humiliated.					
I do not usually hold on to resentment for long.					
When I am under pressure, I maintain self-control.					

Subtotal: ______ / 30

III. Quality of Relationship with the Mother

Item	1	2	3	4	5
I can think about my relationship with my mother without feeling an overwhelming emotional burden.					
I recognize the impact my mother had on my emotional development.					
I have identified wounds, gaps, or patterns derived from that relationship.					
I have worked or am working on forgiveness and healing regarding her.					
My current treatment of women is not dominated by unresolved wounds with my mother.					
I am able to honor what was good and process what was difficult with maturity.					

Subtotal: ______ / 30

IV. Quality of Relationship with Girlfriend or Wife

(If not applicable, answer based on your way of forming emotional bonds.)

Item	1	2	3	4	5
My girlfriend / wife feels emotionally safe with me.					
I listen attentively without reacting defensively right away.					
I consistently express affection, respect, and affirmation.					
I can discuss conflicts without resorting to silence, anger, or manipulation.					
My way of loving combines affection, responsibility, and faithfulness.					
I am willing to correct harmful patterns in the relationship.					

Subtotal: ______ / 30

V. Relationship with Children (Integral Fatherhood)

(Answer according to your reality. If you do not have children, respond based on your potential as a father or your influence on children/young people.)

Item	1	2	3	4	5
My presence in my children's lives conveys security, direction, and stability.					
I am emotionally accessible to my children (they can talk to me without fear).					
I correct with firmness, but without humiliating or hurting.					
I intentionally dedicate time to connect with them, not only to correct them.					
I am an example of character, respect, and self-control before my children.					

In relation to sons

Item	1	2	3	4	5
I am shaping in my son a healthy vision of masculinity.					
I teach him to manage his emotions, not to repress them.					
I model how to treat women with respect and dignity.					
I teach him discipline, responsibility, and purpose.					

In relation to daughters (if applicable)

Item	1	2	3	4	5
My treatment strengthens my daughter's self-esteem and dignity.					

Item	1	2	3	4	5
I am tender without losing firmness or direction.					
My presence communicates protection and emotional security.					
I am an example of how a man should treat a woman.					

Internal Evaluation as a Father

Item	1	2	3	4	5
I parent with intention, not just reacting in the moment.					
I recognize mistakes in my fatherhood and am willing to correct them.					
I am shaping character, not just behavior.					
My fatherhood is aligned with values and purpose, not just culture.					

Subtotal: ______ / 85

VI. Communication Skills

Item	1	2	3	4	5
I know how to listen without interrupting.					
I can express disagreement without offending.					
I ask questions to better understand others.					
My tone of voice and body language usually convey respect.					
I know how to apologize when I am wrong.					
I am able to discuss sensitive topics with maturity.					

Subtotal: ______ / 30

VII. Sexuality and Integrity

Item	1	2	3	4	5
I understand my sexuality as an area that must be governed with wisdom and self-control.					
I do not allow my sexual impulses to control my decisions.					
My treatment of women is not marked by objectification or emotional use.					
I practice healthy boundaries with content, conversations, or environments that weaken my integrity.					
I am honest with myself regarding my sexual struggles.					
My sexual behavior is aligned with my values and principles.					

Subtotal: ______ / 30

VIII. Money, Responsibility, and Stewardship

Item	1	2	3	4	5
I manage money with responsibility and planning.					
My financial decisions are not impulsive.					
I am transparent and trustworthy in financial matters.					
I do not use money to control, impress, or manipulate.					
I know how to distinguish between need, desire, and compulsion.					
My financial management reflects character and a vision for the future.					

Subtotal: ______ / 30

IX. Character, Purpose, and Personal Discipline

Item	1	2	3	4	5
I strive to live by principles, not just by momentary emotions.					
I keep my word to the best of my ability.					
I am disciplined in my personal habits.					
I take responsibility for my actions without always blaming others.					
I have a clear sense of purpose as a man.					
I am committed to growing, healing, and maturing.					

Subtotal: ______ / 30

X. Overall Reading of Your Significant Relationships

Mark how you perceive today the real quality of your most important relationships:

Relationship	Very Weak	Weak	Unsta-ble	Heal-thy	Strong
With my mother	1	2	3	4	5
With my wife / partner	1	2	3	4	5
With my children	1	2	3	4	5
With myself	1	2	3	4	5
With God / spiritual purpose	1	2	3	4	5

Results Sheet

Score by Arca

Area Assessed	Score Obtained	Maximum Score
I. Self-Perception and Identity	____	30
II. Emotional Management	____	30
III. Relationship with Mother	____	30
IV. Relationship with Wife/Partner	____	30
V. Relationship with Children	____	85
VI. Communication Skills	____	30
VII. Sexuality and Integrity	____	30
VIII. Money and Responsibility	____	30
IX. Character, Purpose, and Discipline	____	30

Grand Total: ______ / 325

Interpretation Table

Range	General Reading
266–325	Very strong emotional development. There is visible maturity, although it is still advisable to identify subtle areas for growth.
216–265	Good level of development. There are clear strengths, but some areas require intentional attention.
189–215	Functional level, but with vulnerable areas. It is advisable to work seriously on habits, relationships, and inner healing.
162–188	Important signs of emotional and relational fragility. Intentional intervention, mentoring, or guidance is recommended.
135–161	Low level of emotional development. There are wounds, patterns, or deficits that are likely affecting multiple areas of life.
Less than 135	High-risk emotional and relational condition. It is important to seek help, counseling, or a serious restoration process. If you are considering starting a romantic relationship and choose to first work on these weak areas, you may significantly increase your chances of success.

Reading by Individual Area

Score per Area (out of 30)	Interpretation
26–30	Evident strength
21–25	Good development
16–20	Vulnerable area
11–15	Weak area that requires work
6–10	Critical area

Final Reflection Questions

1. What area of this assessment most clearly reveals my current need?

 ____

 ____

2. What pattern do I see repeating across different areas of my life?

 ____

 ____

3. Which significant relationship needs urgent attention?

 ____

 ____

4. What wound, fear, or habit may be holding back my growth?

 ____

 ____

5. What concrete decision do I need to make starting today?

 ____

 ____

6. What is my expectation as I begin the Emotionally Fit journey?

 ____

____

Initial Action Plan

My three priority areas

1. ____

2. ____

3. ____

Concrete step I will take this week

Emotional habit I need to begin training

____

____

____

____

To Remember...

Being emotionally fit does not mean having no wounds, struggles, or weaknesses. It means recognizing them with courage, taking responsibility for your own growth, and training your heart, mind, and character to become the man God has called you to be.

My courageous resolution:

"I am not a finished man. I am a man in training, and today I choose to grow. Let's get to work!"

CHAPTER 1

LEND ME YOUR LENS

Man. Few words, with so few letters, carry such a broad meaning and generate such far-reaching consequences. Sought after and rejected. Loved and hated. Often misunderstood and, frequently, unfairly judged by those who observe him with reservation. He is called brother, husband, father, friend. You look at him and ask yourself: Who is he?

In this introductory chapter for this journey of emotional growth, let us borrow some lenses so we may clearly see certain realities and dimensions hidden from the human eye. Without a doubt, before going deeper into this subject, we need to understand who we are talking about and what lies behind that masculine figure.

Using the Biologist's Lens

Let us now ask the biologist to lend us his lens so we can better understand what a man is. After reflecting for a moment, he speaks to us about his genetic composition: the male possesses 46 chromosomes, organized into 23 pairs. Of these, the sex pair is XY, and the Y chromosome contains the SRY gene, whose function is to ensure that, during the embryonic process, male differentiation occurs[1] —such as the formation of the testes, the production of testosterone, and the development of characteristics that distinguish the male.[2]

The testes are responsible for producing 95% of testosterone, which is the primary hormone in men.[3] This hormone is responsible for the development of male genitalia, bone density, increased muscle mass, libido (sexual desire), competitive drive, as well as a tendency toward rapid action in stressful or threatening situations.

Why are these behavioral characteristics observed more frequently in men than in women? This is mainly due to genetic and biological factors, as men produce between seven to ten times more testosterone than women.[4]

There are also other very important hormones in men: vasopressin, related to feeling territorial, protection, and bonding; cortisol, associated with the stress response and notably activated in threatening situations; dopamine, linked to reward, motivation, and goal achievement. And, of course, adrenaline and noradrenaline are added, with their well-known fight-or-flight responses.[5]

Men generally have greater muscle mass and strength in their upper limbs. Therefore, when something heavy needs to be moved or carried, it is common to hear women ask, "Where are the men here?"

Let us now use those same lenses to take a closer look at the neurobiological aspect of masculinity. Let us briefly observe the male brain in order to appreciate certain characteristics and better understand how it functions. The neurological organization of the male brain reveals a tendency toward greater intrahemispheric connectivity.[6] This means that men generally feel more comfortable focusing on one task at a time, since their connections occur more within the same hemisphere rather than between both.

This is not the case with women, whose interhemispheric connectivity allows them to perform multiple functions simultaneously without great difficulty.[7] Men, in general, tend to focus on their task, and any parallel assignment may be perceived as a distraction. When it comes to solving a delicate situation, they go straight to the point and are often not ready to immediately connect emotion with action.

This helps explain why women show greater ease in simultaneously combining thoughts, emotions, and the language they use—whether verbal or nonverbal. Does this mean that one is better or worse than the other? In reality, no. The difference is neither qualitative nor hierarchical. What we are observing is a biological difference that deserves to be celebrated as a foundation for complementarity and mutual fulfillment.

Another important aspect that the biologist's lens offers regarding masculinity relates to the reproductive system. The testes function as a factory that constantly produces sperm. In men, hormonal production does not occur in cycles but rather continuously; therefore, they can potentially maintain reproductive capacity well into later adulthood.

Regarding longevity, a large body of data suggests that there is a difference in life expectancy between men and women. The global average is currently estimated at around five years, [8] although other studies suggest the difference may range between three and ten years.[9] Consistently, men have shown a tendency to live fewer years than women.

It is worth noting that this figure is not fixed, as it may vary depending on the country and certain conditions. For example, men tend to visit the doctor less frequently.[10] Some men, even when they feel healthy, have not yet identified a primary care physician. Often, this practice is associated with the late detection of serious health problems that could have been prevented. This helps explain the increase in obesity, high cholesterol, and hypertension in this population, which compromises cardiovascular health.[11]

From Man to Man

Based on what we have seen above, what would be three strengths of men?

________________________________,

________________________________,

________________________________.

If a man could improve two of the aspects mentioned above, what would they be and how would this help him in life?

________________________________,

________________________________,

________________________________.

Take a few minutes to answer honestly. These questions are for your personal growth.

Through the Lens of a Sociologist

The sociologist's lens quietly observes that the role of the male figure has undergone changes over time. Initially, he is the object of provision. Later, he becomes responsible for providing for his family. Eventually, he becomes a co-provider and shares with his wife the care of their children. His presence—both

physical and emotional—becomes not only necessary but vital, as the father is a mirror through which a child can see himself reflected[12] and the model he may one day surpass.

Positively or negatively, a man influences his environment and is, in turn, affected by the dynamics around him. The continuous and disproportionate rise in the cost of living, lack of job security, and the pressure of workplace competitiveness are just some of the socioeconomic realities that challenge a man's peace. Unfortunately, for most men, what they achieve and what they provide becomes tied to their personal worth, which can negatively impact vital organs of the body.

On the other hand, today it is striking that men are experiencing a concerning level of digital loneliness. While technological development has made virtual interactions more common, we are also seeing a man who is less socially connected.[13] Studies highlight that overexposure to digital media may be linked to loneliness, anxiety, stress, and certain health issues among men. Another recent study found a positive relationship between loneliness and pornography use.[14] For this reason, it is effective to promote the cultivation of meaningful friendships that add value to your life and help you become a better person. These support networks are invaluable in times of emotional need.

Likewise, in some cases, conditioning and training—along with one's family pattern, if applicable—serve as instructions for shaping a style oriented toward aggression. These characteristics of hegemonic masculinity should be viewed as a model that has generated divisions and trauma.

My Personal Reflections

When I think that the father is the mirror through which a child learns to see himself, what images or ideas come to mind?

____,

____,

What social skills would I like to develop to expand and strengthen a circle of meaningful friendships?

____,

____,

Seeing Man Through the Lens of a Psychologist

Hegemonic masculinity is marked by a strong cultural framework that, even today, continues to be accepted or considered the norm in many places. From a young age, boys have been taught that a man "has to" and "must" "be strong." In this context, expressing emotions or being expressive is seen as a sign of weakness. Thus, the phrase "men don't cry" has become widely accepted. It seems that society expects men to limit the expression of their feelings and even the desire to cry, no matter what happens. Is man truly a being without emotions? Or has he, in many cases, been taught to repress them?

Based on the pillars of this framework, tasks that belong to any human being are viewed as specifically feminine. The pursuit of power, money, and sex (at almost any cost) is strongly promoted, to the point that excelling in these areas becomes, for some, not only a source of pride but also a mark of social superiority.[15]

From man to man, we ask: could this have repercussions on a personal, family, and social level?

Can men suffer from anxiety? Of course they can. However, it is often expressed in terms of anger, irritability, or withdrawal. These realities help explain the high rates of depression and male suicide. A recent report revealed that in the United States, male suicide rates are 3.8 times higher than those of women. [16] Globally, it is estimated that men double the suicide rates of women.[17]

At times, male mental health may present what is known as the "exhausted man" syndrome (also known as burnout). Here we see a man with chronic exhaustion, stress caused by multiple financial commitments and limited coping capacity, complex work conditions that generate pressure, and an overload of demands constantly weighing on his shoulders.

What about the expectations placed on men? The competitive environment surrounding them raises expectations within their gender, and these tend to be even higher from the opposite sex. One day, a man approached me and said with frustration: "How I wish I were as handsome as my mother says I am, as strong as my children believe I am, and as rich as my wife seems to think I am!" The contrast between expectations and reality causes some men to feel unable to measure up, leading them at

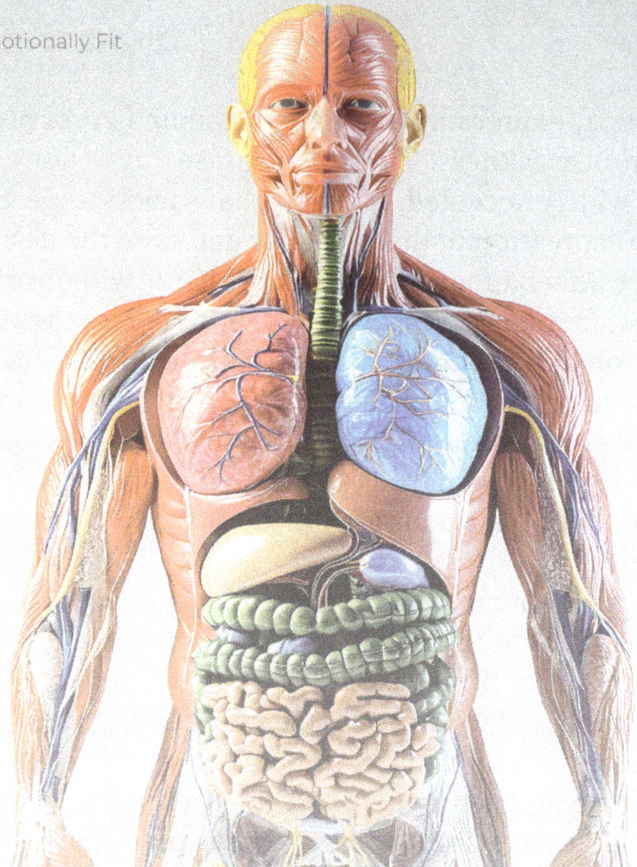

times to exaggerate in order to cover the gap between what they think they are (or what others expect them to be) and what they truly are. It must be acknowledged that, from an emotional standpoint, this effort is often free—but not cheap. Today, male identity, self-worth, and sense of purpose are under constant threat.

If we continue using the psychologist's lens, we cannot ignore realities that were once very real but today remain silent. No one talks about them, yet silence does not erase the truth of childhood wounds—unresolved issues with a parent, abandonment, roots of bitterness due to unequal treatment at home, shame over something we did not know how to express, and the deep desire for recognition that was never fulfilled.

Expressing our vulnerabilities does not make us less of a man. For our own good, and for the sake of the next generation of men who look up to us as models, we must create a safe space for healthy masculinity. We must understand that our wounds can be healed in Christ. And if help is needed, why not seek it? We can give ourselves that well-deserved permission!

Let Us Consult the Bible and Complete the Idea

Sometimes we feel tired and wonder if what we are doing is truly worthwhile. The Bible has a message of hope for you and for every man. Look up the verse and complete the idea:

1. "The Lord is near to the __________ of __________; and saves those who are __________ in __________." Psalm 34:18
2. "Come to __________ all who are __________ and __________, and I will give you __________. Take my yoke upon __________, and __________ from me, for I am gentle and __________ in heart, and you will find __________ for your souls." Matthew 11:28–29
3. "He heals the __________ of heart and binds up their __________." Psalm 147:3
4. "Do not __________, for I am __________; do not be __________, for I am your __________ who __________ you; I will __________ you, I will __________ you with the right hand of my __________." Isaiah 41:10
5. "Casting all your __________ upon Him, because He __________ for __________." 1 Peter 5:7

Activity: "The Invisible Backpack" (The Exhausted Man)

Objective: To make mental and emotional overload visible.

Instructions:

- Each participant writes on cards:
  - responsibilities
  - financial concerns
  - external expectations
- These are symbolically placed in a backpack.

Key message:

"Not every burden is bad, but carrying it alone is."

Reflection:

What burden is not mine to carry? ____________________

What burden do I need to share? ____________________

Seeing Man Through God's Lens

With great anticipation, we consider what it would mean for God to lend us His lens to see the man He Himself created. The biologist's lens impresses us with its impact, the sociologist's lens broadens our understanding, the psychologist's lens unsettles us somewhat—but we grow still as we prepare to put on the Creator's lens.

Before the climactic moment, God had created all natural beauty with the breath of His mouth. Psalm 33:6 declares: "By the word of the Lord the heavens were made, and all their host by the breath of His mouth." Verse 9 adds: "For He spoke, and it was done; He commanded, and it stood fast." Thus, we see a world full of unimaginable wonders, functioning in perfect order.

God was about to create His masterpiece—and He was not satisfied with words alone. So He chose to become personally involved. On the sixth day, God said: "Let Us make man in Our image, according to Our likeness; let them have dominion..." (Genesis 1:26). The brilliant mind that created the universe now pauses to design, with His own hands, the most complex creation: the human being.

Just to have an idea of the magnitude of the project, look at how many systems compose it and the main reasons that motivated their existence:

- **Nervous** – Central command of the body's functions.
- **Skeletal** – It is estimated that an adult human being has 206 bones[18] that provide support to the body and protect its vital organs.
- **Endocrine** – Responsible for hormonal regulation for the proper functioning of bodily processes that facilitate growth, metabolism, and reproduction.
- **Digestive** – Takes the food we ingest, absorbs its nutrients, and discards what it does not need.
- **Respiratory** – Allows man to breathe and to exchange oxygen and carbon dioxide.
- **Urinary** – It is the excretory system. Its function is to eliminate liquid waste and maintain the balance between water and electrolytes.
- **Circulatory** – Responsible for the cardiovascular aspect. It transports blood, oxygen, nutrients, and hormones.
- **Muscular** – The human body has more than 600 voluntary and involuntary muscles[19] that facilitate posture, movement, and heat production.
- **Reproductive** – Designed to guarantee reproduction and continuity of the species.

- **Integumentary** - Designed to protect your body. It includes nails, hair, skin, and sweat glands.
- **Immune** - Conceived to defend the body against infections and diseases.

Does it not seem astonishing to you the work of God and the precise order He followed to care for His special creature? Definitely, every detail set into action not only causes admiration, but speaks loudly of a God who is foresighted, a provider, and loving.

It draws attention to the fact that each system of the human body was created with a purpose. The way in which these systems work so harmoniously and cooperate with one another to promote the full well-being of the individual gives us a faint reflection of an even greater plan that God had in mind when creating us. So many details in planning and such precision in execution speak to us of a purpose and a mission that stand out. The biblical text declares in Genesis 1:27 that "God created man in His image." His plan was that "the longer man lived, the more fully he would reveal that image and reflect the glory of the Creator,"[20] who said: "All who are called by My name; for My glory I have created them, I have formed them and made them" (Isaiah 43:7).

The Creator is a God of relationships. He likes to share with His creatures, whom He loves so much. For this reason, before creating Eve, Adam had a space not only to name the animals, but to share with his Maker and establish a spiritual bond with Him. That bond with God would be a key element for man to fulfill his purpose.

From Genesis 3 onward, the entire picture changes after the entrance of sin. With the fall, the male role ends up being distorted. By disconnecting from God, a chain of conflicts began—fear, guilt, shame, escape, and aggressiveness. Genesis 3:15 provides the announcement of a Savior, bringing a light of hope to this family in crisis. The Son of God would give Himself to save man from sin and, in due time, restore his relationship with Him, his life, and his family.

Masculinity finds its model par excellence in Christ, who is firm, yet at the same time tender. He is powerful and also compassionate. He is just and also merciful. His leadership is strong, but it wins the heart through service and self-giving. This model of masculinity is not defined by the temporary standards of today's society, but by the timeless principles that make the government of Christ unique.

In a postmodern world like the one in which we live, in our condition as men, we all have our pressures. They may be linked to work, family, finances, health, and social expectations themselves. There is always someone watching to ensure that the man behaves according to the social standard. It is there when each man needs to remember that he was created with a great purpose and that, with the help of God, he can live his masculinity with dignity and excellence. In the following chapters, we will explore this path.

Group Reflection

1. If I was created in God's image and with purpose, what does that mean for my identity as a man?
2. In light of that purpose, what concrete changes do I need to make?

____

____

Training My Brain

"A Concrete Commitment"

Objective: Move from reflection to action.

Instructions:
Complete the following:

- This week I commit to...

  ____

- To care for my mind, health, and heart, I will...

  ____

- To strengthen my connection with God, I will focus on...

  ____

Endnotes:

1. Sekido R, Lovell-Badge R. Sex determination involves synergistic action of SRY and SF1 on a specific Sox9 enhancer. Nature. 2008 Jun 12;453(7197):930-4. doi: 10.1038/nature06944. Epub 2008 May 4. Erratum in: Nature. 2008 Dec 11;456(7223):824. PMID: 18454134.

2. Sadler, T. W. (2019). Langman's Medical Embryology. Wolters Kluwer.

3. Melmed, S., Auchus, R. J., Goldfine, A. B., Koenig, R. J., & Rosen, C. J. (2020). Williams Textbook of Endocrinology (14th ed.). Elsevier.

4. Guyton, A. C., & Hall, J. E. (2021). Textbook of Medical Physiology (14th ed.). Elsevier.

5. Expanded by Guyton, cited above.

6. Ingalhalikar M, Smith A, Parker D, Satterthwaite TD, Elliott MA, Ruparel K, Hakonarson H, Gur RE, Gur RC, Verma R. Sex differences in the structural connectome of the human brain. Proc Natl Acad Sci U S A. 2014 Jan 14;111(2):823-8. doi: 10.1073/pnas.1316909110. Epub 2013 Dec 2. PMID: 24297904; PMCID: PMC3896179.

7. Kenia Martínez, Joost Janssen, José Ángel Pineda-Pardo, Susanna Carmona, Francisco Javier Román, Yasser Alemán-Gómez, David Garcia-Garcia, Sergio Escorial, María Ángeles Quiroga, Emiliano Santarnecchi, Francisco Javier Navas-Sánchez, Manuel Desco, Celso Arango, Roberto Colom, Individual differences in the dominance of interhemispheric connections predict cognitive ability beyond sex and brain size, NeuroImage, Volume 155, 2017, Pages 234-244, ISSN 1053-8119, https://doi.org/10.1016/j.neuroimage.2017.04.029.

8. Wu, Y.-T., Niubo, A. S., Daskalopoulou, C., Moreno-Agostino, D., Stefler, D., Bobak, M., Oram, S., Prince, M., & Prina, M. (2021). Sex differences in mortality: results from a population-based study of 12 longitudinal cohorts. Canadian Medical Association Journal, 193(11). https://doi.org/10.1503/CMAJ.200484

9. OECD (2025), Health at a Glance 2025: OECD Indicators, OECD Publishing, Paris, https://doi.org/10.1787/8f9e3f98-en.

10. Wang Y, Hunt K, Nazareth I, Freemantle N, Petersen I. Do men consult less than women? An analysis of routinely collected UK general practice data. BMJ Open. 2013 Aug 19;3(8):e003320. doi: 10.1136/bmjopen-2013-003320. PMID: 23959757; PMCID: PMC3753483.

11. Framingham Heart Study. (s. f.). Coronary Heart Disease (10-year risk). Framingham Heart Study. https://www.framinghamheartstudy.org/fhs-risk-functions/coronary-heart-disease-10-year-risk/

12. Drescher, J. M. (1996). Siete necesidades básicas del niño. Mundo Hispano.

13. Cherepiekhina, O. A. (2025). Digital loneliness and men's health: An interdisciplinary approach (systematic review) [Preprint]. ResearchGate. https://www.researchgate.net/publication/390519861_O_A_Cherepiekhina_Digital_loneliness_and_men%27s_health_an_interdisciplinary_approach_systematic_review

14. Vescan M, Flack M, Caudwell KM. Loneliness and problematic pornography Use: What is the role of emotion regulation and interaction with content Creators? Addict Behav Rep. 2024 May 3;19:100550. doi: 10.1016/j.abrep.2024.100550. PMID: 38746953; PMCID: PMC11090905.

15. For more details on this point, see Chapter 1 on the male gender in the book Spark Up Your Marriage, by the same author.

16. American Foundation for Suicide Prevention. (2025). Suicide statistics. https://afsp.org/suicide-statistics/?utm.com

17. International Association for Suicide Prevention. (2021. WSPD facts & figures infographic [Infografía]. https://www.iasp.info/wp-content/uploads/WSPD-Facts-Figures-Infographic.pdf

18. Standring, S. (Ed.). (2021). Gray's Anatomy: The Anatomical Basis of Clinical Practice (42nd ed.). Elsevier.

19. Moore, K. L., Dalley, A. F., & Agur, A. M. R. (2023). Clinically Oriented Anatomy (9th ed.). Wolters Kluwer.

20. White, E. G. (2013). Patriarcas y profetas (pp. 45–46). Asociación Casa Editora Sudamericana.

CHAPTER 2

IDENTITY AND PURPOSE

There are two fundamental questions on the stage of life that every man must ask himself at some point: Who am I? and What am I doing here? Although one may help answer the other, both are essentially different. One relates to identity. The other, to purpose.

The healthy and successful development of masculinity largely depends on each man's ability to understand and live according to what these questions imply. The crises observed in multiple areas of life, suggest that many men are still questioning themselves in their search for answers to these fundamental questions.

Outcomes Connected to Identity Crisis

Why is the issue of male identity important, and what risks are assumed when its relevance is ignored? While we cannot claim that the situations presented below are the direct and exclusive result of a lack of identity, there is no doubt that, in one way or another, they may be connected to it.

To illustrate some of these realities that have become so common today, the Centers for Disease Control and Prevention (CDC) reported in April 2025 that during the year 2023, male suicide was nearly four times higher than female suicide. During that period, men accounted for approximately 80% of all suicide deaths in that country.[1]

Likewise, referring to reports from the 2021–2023 period, the same government agency indicated that, of the 105,007 overdose deaths,[2] men led the statistics with alarming figures: 44.3 per 100,000, compared to 18.3 among women.[3]

What do these numbers tell us? They tell us that, like every human being, men have emotional needs. They tell us that a man does not become weaker by expressing what he feels and that, when needed, asking for help is not a sign of weakness. When a man does not have a clear sense of who he is, in complex and emotionally intense moments, he may easily identify with silence, isolation, and loss of hope. And if we look closely, those three elements are fuel for suicide.

These data also tells me that men go through moments of deep emotional pain, even though many never verbalize it. I have heard influential voices suggest that men should not express their emotions and should hide their internal struggles. In the interest of intellectual honesty, I would have also liked to hear them acknowledge that substance use, work addiction, unnecessary risk-taking, and isolation are not reliable anesthetics for emotional repression.

And what about adolescent boys? Some think they do not suffer, do not feel sadness, and have never experienced discouragement. The fact that statistics show higher numbers

in women does not mean that men are actually better off. The social conditioning many have received almost penalizes men for opening the "box" of their hearts and revealing what is inside.

For this reason, the emotional complaints of many men will rarely appear directly in statistics. We must learn to read their reports through equivalent behaviors: constant irritability, high-risk behaviors, physical and emotional distancing, prolonged silence, and addiction-related problems. In light of this reality, the question remains valid: what is the advantage, if any, of continuing to ignore male identity?

Difficulties associated with male identity crisis are often perceived as minimal or insignificant. Although this perception may seem like a personal choice, it is often driven in part by the false sense of "I am in control," "I can handle it on my own," or the familiar phrase: "No, there is no problem... everything is fine..." These ideas create a temporary sense of calm that, in practice, ends up justifying social isolation.

That disconnection has an extraordinary impact. According to conclusions from the Office of the Surgeon General of the United States, loneliness can increase the risk of death by 26%, and social isolation by 29%. Likewise, the magnitude of this type of disconnection has been compared to the damage caused by smoking approximately 15 cigarettes per day. Added to this are higher probabilities of depression, anxiety, and cardiovascular disease.[4] It is important to note that a disconnected man is not necessarily one who lacks a network of friends, but also one who refuses to ask for help or seek timely counsel when needed.

On the other hand, men with a limited sense of personal identity often prefer to remain single, even when they express interest in forming a family. Research indicates that in the United States, 63% of men age 30 or younger consider themselves single.[5] Meanwhile, their sense of commitment is not strongly stimulated to grow, the relationships they form tend to be fragile, the risks of disordered sexuality increase, and their family project is postponed.

Likewise, the lack of male identity contributes to father absence, which by 2023 affected approximately 25% of children.[6] Relevant studies have found a relationship between absent fathers and school-age children and young adults who present higher levels of depression,[7] socio-emotional difficulties, and problems in the educational environment.[8]

Group Discussion

1. In a man's life, there are areas where we need support, guidance, or counsel; however, we prefer to remain silent—even when we have people nearby who could help us. What are some of those areas in your personal experience?

    ____

    ____

2. Choose one of the areas mentioned and reflect: How could things change if you had adequate, timely, and trustworthy mentorship in that area?

    ____

    ____

3. What concrete step am I willing to take in the coming days to stop facing one of these areas alone and begin walking with healthy mentorship?

____

____

The Sense of Identity

We can define the sense of identity as the inner and constant perception that allows you to clearly know who you are, how much you are worth, why you are here, and where you are headed. A man with a healthy sense of identity can be at peace with himself because he understands that his value does not depend on his level of productivity. He sees himself and describes himself for who he truly is, without allowing himself to be framed by others' perceptions. He appreciates the fact that he is here, understands why he is here, and moves forward with expectation toward the future ahead of him.

To Clarify...

- **Profession or productivity does not define who you are.** What you do is one thing; who you are is something entirely different. One day you will stop doing what you enjoy and what many have associated you with, but that day you will not cease to exist. You will still be there, and the value God gave you will not have changed at all.
- **Position or possession does not define who you are either.** Just as life is temporary, so are positions and possessions. Therefore, the role you hold today, your current social status, and the assets that make up your portfolio are insufficient to determine who you are or how much you are worth as a person. Building them may take a lifetime—or even generations—but everything can vanish with something as simple as a bad decision. And then what?

- **Others' perceptions are not always reliable**. Social media often portrays a character; who we truly are is something else. The level of social approval does not define who we are either. Entering the emotional dance—joy or sadness—based solely on this can be dangerous. It is like trading a gold bar for a piece of bread.
- **The opinion of others**—do you think it is worthy of full credit, especially when it comes to something as important as defining who you are and how much you are worth?

The man who builds his value on the based on his job; who rises only when he feels successful and perceives failure as a grave; who collapses under criticism; and upon noticing that his body is aging, believes that everything is over—will inevitably struggle with confusion, anxiety, and a deep sense of existential emptiness. Externally he may appear functional, but internally he is empty.

On the contrary, when a man understands that his value was assigned to him when he was created by the God of the universe; when he accepts that this value was fully confirmed on the day Christ paid the full price for his redemption on the cross of Calvary—at the highest cost ever imagined—that day the problem of his identity and value is resolved.

Many people will be confused and will not know who you are. Some will deliberately ignore you. Others will try to assign you a value made of crumbs and leftovers. Stay calm. If that is the best they have, there is no reason to react violently to their unfortunate ignorance. In that moment, lift your eyes to heaven and, by faith, contemplate the One who said:

"For you were precious in My sight, you have been honored, and I have loved you; therefore I will give men for you, and people for your life." (Isaiah 43:4)

You may wonder why the infinite God has been willing to do so much for you. His answer is simple: because He loves you. In

fact, He declares: “I have loved you with an everlasting love...” (Jeremiah 31:3).

Why Am I Here?

If we consider the data presented at the beginning of this chapter regarding outcomes associated with male identity crisis, we realize that this is not an optional issue for a man’s physical, mental, and spiritual health. Men who clearly understand who they are, why they are here, and where they are going show lower rates of premature mortality, lower levels of depression, and greater resilience in the face of life’s adversities. As a result, they provide their children with healthy modeling that prepares them for a brighter future.

What does a man live for? A wise answer to this question gives direction beyond his occupation. His reason for existing will define his goals, his challenges, the priority use of his energy, and where he places the weight of his responsibility. Questions arise such as: What will I fight for? What do I want to build? And if at the end of everything I could preserve only one thing, what would it be?

In seeking the true purpose of life, we can turn either to the creatures that inhabit the earth or to the Creator of heaven and earth. Creatures may offer elaborate philosophical and ontological arguments that may, to some degree, appeal to human intellect. However, if we ask the Creator of man, He can tell us why He made him. The Bible, also known as the Holy Scriptures, reveals great mysteries about God and answers specific questions such as this one.

Let Us Consult the Bible and Complete the Idea

1. Isaiah 43:7 tells us: "All who are called by My name, whom I have created for ____________ ____________ ____________, I have formed him, yes, I have made him."
2. 1 Corinthians 10:31 reminds us: "Therefore, whether you eat or drink, or whatever you do, do all ____________ ____________ ____________ ____________ ____________."
3. If the meaning of life is found when a man chooses to live not for himself but for the glory of God, then what would happen if we sought to know God better, understand His plan more clearly, and learn to live according to His purpose? A man's highest good is far from merely "appearing." It lies in "being," while shaping a character for eternity.

My Personal Reflection Exercise

1. If my life ended today:
   a. Whose agenda would I have been following? ____________
   ____

b. What purposes would I have liked to include to live life more fully? ________________, ________________, ________________, ________________, ________________

c. What concrete step would I like to take this week to make the most of the opportunities God has placed before me? ________________, ________________, ________________, ________________, ________________

From Man to Man

Discuss as a group the idea: If my identity as a man is hidden in Christ and the meaning of life is best understood by living for His glory, what specific changes could this bring...

1. To my life: ____________, ____________, ____________.
2. To my family: ____________, ____________, ____________.
3. To my social interactions: ____________, ____________, ____________, ____________.

How to Train the Brain to Strengthen My Sense of Identity

When a man comes to understand who he is and discovers his true value, his life can change forever. Therefore, defining and strengthening one's sense of identity is a vital matter for every man. It is not an act completed in an instant, but a process of training that integrates several aspects. To facilitate this, we will include principles from psychology, neuroscience, and theology, which can operate in a complementary way.

Why these three elements? Psychology helps us understand, in practical terms, who I am. Neuroscience explains how the sense of identity is formed and how it can be changed in the brain. Theology provides an admirable definition that answers two essential questions: why I exist and before whom am I accountable.

Every person has a life story, also known as their personal narrative. This internal story not only integrates the past but also the present and the future, providing a sense of coherence that gives unity and purpose to life.[9]

Practical Training

Exercise 1 – The Male "Self" Narrative

I will write my story in three chapters:

Chapter 1: Where do I come from?
Chapter 2: What valuable lessons have I learned so far?
Chapter 3: Where am I headed? or Where do I want to go?

Exercise 2: Clarity of Roles and Values

I want to strengthen my identity and, at the same time, avoid confusion between my roles and my values as a result of social pressure. Therefore, this week I will focus on the following training:

- I will make a list of my roles by category: son, brother, husband, father, employee (or employer), believer (if I am one).
- In each role, I will write one behavior that is non-negotiable.

Example: As a son: Show respect to my parents
As a husband: Love my wife

Exercise 3: Developing a Mature Masculine Identity

A man strengthens his psychological stability when he develops autonomy, a sense of competence, and relational capacity.[10] This way, his response will not be based on the level of approval he receives, but on the conviction he has developed based on values and principles.

a. Autonomy

I know that sometimes I act under pressure. Instead, these are the things I choose to do out of conviction:

b. Competence

What skill or skills should I train this month?

c. Relationships

Which relationships should I seek to care for or restore at this time?

What concrete things can I do this week to move in that direction?

Relationship: ____________________

Things I can do this week: ____________________

____________________ ____________________

Exercise 4: Model of Masculinity

Cultural norms often define a model of masculinity centered on dominance, self-sufficiency, rigid expectations, and situational emotional management.[11]

a. Evaluate the cost of these traditional behaviors:

b. Create two columns:

In one column: behaviors of a man that come from the model of Christ and the values promoted by His teaching.

In the other column: behaviors that are the result of cultural or social pressure.

c. This week I will be intentional in replacing "harshness" or "toughness" with tenderness and gentleness. When the desire to dominate arises, I will seek to replace it with servant leadership.

Exercise 5: Habits — Automation

Studies have shown that habit automation can increase significantly with practice.[12] It is important to note that results are not magical or immediate and may vary from person to person.

a. This week we will work on identity habits. They do not have to be big. It is enough that they are small, daily, and measurable.

Examples:

- "Each day I will dedicate 10 minutes to Bible study."
- "Before going to sleep, I will write a short paragraph summarizing the lessons I learned during the day."
- "I am choosing to perform an act of service or show respect intentionally every day."

b. Connect the habit you want to develop with an identity:

- "I have decided to study the Bible because I am a God-fearing man."
- "I do this because I am an emotionally responsible man."

c. Reduce friction through consistency:

- Do it at the same time, if possible in the same place, and following the same order.

d. Prepare your environment so that habit formation becomes easier over time.

- It may seem difficult at first, but once strengthened, the habit will become part of your life.

e. Focus on consistency.

- A small action repeated many times is more powerful than a great burst of occasional motivation.

- To establish and strengthen a habit, 5 minutes daily for six months is more fruitful than 60 minutes once a week for a single month.
- Consistency is the language the brain interprets as "security,"[13] which favors the consolidation of neural circuits.[14]

If You Have a Son...

If you have a son, helping him build his identity is a matter of great importance. A key question is: *How does he see his father?* Researchers affirm that a son's perception of his father varies depending on his stage of development:

- **In the case of a baby,** his father is seen as his base and emotional refuge. For him, the word "dad" is synonymous with security.[15]
- **For a child**, his father is his hero, his model of masculinity. [16] From him, he learns what it means to be a man and how women should be treated. The child learns primarily through observation and then internalizes what he sees, assimilating it as good and valid.

- **The adolescent** sees his father as both a mirror and a contrast. The father becomes his frame of reference as he searches for his own identity. Although he may criticize his father, deep down he admires him, compares himself to him, and he continues to be a powerful influence in his life.[17]
- **When he becomes a man**, the son sees his father from another perspective. Now both his strengths and his wounds are revealed. The father remains the root, but is no longer just an authority figure, but also an advisor. At this stage, understanding the father's story facilitates forgiveness, strengthens the relationship, establishes healthy boundaries, and allows differentiation.[18]

Helping a son transition from childhood to responsible adulthood is not an easy task. Issues related to identity, sense of purpose, and the progressive management of adult responsibilities often represent a real challenge. Among these are a set of skills and knowledge he should ideally possess by the age of 18. These include, among others: personal identification, work, finances, digital safety, home safety, cleaning, waste management, basic knowledge of electricity, plumbing, heating, air conditioning, gardening, vehicle maintenance, and travel.

Recently, I spoke with my friend Marty Miller from Blueprint, a nonprofit organization based in Tennessee, United States, dedicated to promoting the holistic development of masculinity. What he shared with me on this subject seemed valuable and highly relevant for every father raising a son. If you wish to review the full document, you may look for the reference at the bottom of the page and visit their website to access the article for free.[1]

My Personal Reflections

____

____

____

____

____

Endnotes:

1. Centers for Disease Control and Prevention. (2025. Suicide data and statistics. https://www.cdc.gov/suicide/facts/data.html?.com

2. Centers for Disease Control and Prevention. (2024). Trends in state suicide rates — United States, 1999–2021 (Data Brief No. 522). National Center for Health Statistics. https://www.cdc.gov/nchs/products/databriefs/db522.htm?.com

3. Ibid.

4. Office of the Surgeon General. (2023). *Our epidemic of loneliness and isolation: The U.S. Surgeon General's Advisory on the healing effects of social connection and community* (U.S. Department of Health and Human Services). https://www.hhs.gov/sites/default/files/surgeon-general-social-connection-advisory.pdf

5. Pew Research Center. (2023, febrero 8). *For Valentine's Day, 5 facts about single Americans.* https://www.pewresearch.org/short-reads/2023/02/08/for-valentines-day-5-facts-about-single-americans/?.com

6. Chamie, J. (2025, enero 13). *America's single-parent households and missing fathers.* N-IUSSP. https://www.niussp.org/family-and-households/americas-single-parent-households-and-missing-fathers/.com

6. Culpin I, Heuvelman H, Rai D, Pearson RM, Joinson C, Heron J, Evans J, Kwong ASF. Father absence and trajectories of offspring mental health across adolescence and young adulthood: Findings from a UK-birth cohort. J Affect Disord. 2022 Oct 1;314:150-159. doi: 10.1016/j.jad.2022.07.016. Epub 2022 Jul 14. PMID: 35842065; PMCID: PMC10666570.

7. McLanahan S, Tach L, Schneider D. The Causal Effects of Father Absence. Annu Rev Sociol. 2013 Jul;39:399-427. doi: 10.1146/annurev-

soc-071312-145704. PMID: 24489431; PMCID: PMC3904543.

8. McAdams, D. P., & McLean, K. C. (2013). Narrative Identity. *Current Directions in Psychological Science, 22*(3), 233-238. https://doi.org/10.1177/0963721413475622

9. Ryan, R. M., & Deci, E. L. (2000). *Self-determination theory and the facilitation of intrinsic motivation, social development, and well-being.* https://selfdeterminationtheory.org/SDT/documents/2000_RyanDeci_SDT.pdf

10. Mahalik, J. R., Locke, B. D., Ludlow, L. H., Diemer, M. A., Scott, R. P. J., Gottfried, M., & Freitas, G. (2003). *Development of the Conformity to Masculine Norms Inventory.* Psychology of Men & Masculinity, 4(1), 3–25.

11. Lally, P., van Jaarsveld, C. H. M., Potts, H. W. W., & Wardle, J. (2010). *How are habits formed: Modelling habit formation in the real world.* **European Journal of Social** Psychology, 40(6), 998–1009. https://doi.org/10.1002/ejsp.674

12. Friston K. The free-energy principle: a unified brain theory? Nat Rev Neurosci. 2010 Feb;11(2):127-38. doi: 10.1038/nrn2787. Epub 2010 Jan 13. PMID: 20068583.

13. Porges, S. W. (2011). *The polyvagal theory: Neurophysiological foundations of emotions, attachment, communication, and self-regulation.* Norton.

14. Bowlby, J. (1988). *A secure base: Parent-child attachment and healthy human development.* Basic Books.

15. Lamb, M. E. (2010). *The role of the father in child development* (5th ed.). Wiley.

16. Allen, J. P., & Tan, J. S. (2016). The multiple facets of attachment in adolescence. *Handbook of Attachment* (3rd ed.). Guilford Press.

17. Main, M., Kaplan, N., & Cassidy, J. (1985). Security in infancy, childhood, and adulthood: A move to the level of representation. *Monographs of The Society for Research in Child Development, 50*, 66-104.

18. To access the full article, visit: https://blueprintformen.org/wp-content/uploads/2023/02/The-Adulting-Challenge-Blueprint-2023.pdf

CHAPTER 3

CONTROLLING THE UNCONTROLLABLE

I met Edward. A man with extraordinary wisdom; tall, strong, masculine. His presence and natural style required no argument to prove that he was a leader admired by many. When his laughter burst forth, it drew the laughter of others, yet his wife and children publicly avoided him. What was happening with this good man? Whatever occurred—small or large—if it did not align with his worldview, it became a reason for a heated argument. His large eyes would widen as if ready to cast fire around him. His mouth, which at times spoke gentle words, would now speak arrogantly, intimidatingly, and without control. His wife, whom he loved deeply, feared him. His children carefully measured their words.

Later, outside the home, these behaviors became more frequent. Edward was now viewed with concern. His episodes of uncontrolled anger escalated in intensity. One day, a friend asked him why he behaved this way. His simple answer was: "That's who I am; that's how I was formed." Then he added: "And let it be clear, I'm much better than my father, okay."

Edward's story repeats itself often. The De Leóns use the term "broken and torn hearts"[1] to describe a condition far more common than we tend to imagine. For our part, the model of masculinity we have received has not always been the best. In many cases, those inherited models deeply wound our emotions. The blood from those wounds still flows, and sometimes drops we cannot even see splash and stain the eyes and faces of those we love.

The Emotional Alarm Center

Some people ask: what happens in a man's brain when he is filled with anger? The human being possesses what we could call an Emotional Alarm Center, known as the amygdala. When activated rapidly and intensely, it can trigger unpleasant reactions. Let me explain.

The prefrontal cortex is part of the frontal lobe and is responsible for executive functions that allow us to evaluate and process facts and information.[2] It uses thinking and analysis to make mature and sensible decisions. The amygdala and the prefrontal cortex work together.[3]

In fact, the way a man manages his emotions, the ability he demonstrates to control anger, regulate stress, and the decisions that result from these processes depend largely on how closely these two structures function together. The prefrontal cortex is the reasoning center that helps regulate and control emotions. Aspects as important as willpower,[4] morality,[5] decision-making,[6] anger control,[7] the ability to empathize,[8] spirituality, [9] and impulse control[1] are rooted in the prefrontal cortex. Does that seem like a lot of responsibility for one part of the brain? Indeed, it is.

It is important to note that the prefrontal cortex operates more slowly than the amygdala. One could say that the amygdala acts before thinking, while the prefrontal cortex produces thought before action, doing so in a planned and strategic way. For this reason, its functioning reflects social behavior marked by empathy and sound values.

By default, the amygdala detects threats and quickly activates emotional responses that are often intense such as: fear, explosive anger, anxiety, guilt, shame, shock or paralysis, flight, hypervigilance, among others. The threat does not have to be real for these reactions to occur; it only needs to be perceived or imagined and not properly processed.

It is also worth noting that the male response to these emotions may differ significantly from that of women. According to the pillars of hegemonic masculinity discussed in this work, during development the male mind is often caught between what is feared and what is desired.

What is desired is defined by: how I want to be seen by others. What is feared is defined by those things from which I prefer to distance myself to avoid being associated with them.

- What do I desire, and what do I fear?
- To be seen as successful or as a failure?
- To be considered brave or cowardly?
- To be perceived as someone who governs and dominates, or as someone who is governed and dominated?[11]

The Contrast in Action

Due to multifactorial elements—including biological, hormonal, cultural, and social learning factors—the intense emotional responses of the amygdala manifest in men in a very particular way. The resulting experiences may differ between a man and a woman.

For example, what does a woman do when she feels fear? Often, she seeks refuge in the arms of a man who provides security—whether a father, brother, husband, or friend. And what does a man do in response to the same emotion? To better understand this, we must identify what lies beneath the male mind.

In the case of many men, from an early age they assumed—based on their models and social environment—that fear is a sign of weakness. Therefore, vulnerability, for those who hold this belief, is something they cannot display without paying a high cost in terms of social perception. Does this mean that men do not feel fear? Emotions have no gender. Therefore, under certain circumstances, anyone can feel fear.

What, then, does a man do when he is afraid? In many cultures, anger is an acceptable emotion, but fear is not. Thus, many men react with anger in response to fear. Their behavior may become aggressive; they appear irritated and easily provoked. Their demeanor becomes rough or harsh, they speak loudly and in a dominant tone, attempting to conceal that they are actually being dominated by fear and circumstances.

How is male anger expressed? Male anger is usually expressed more actively[12] than reflectively[13] and manifests in a physical and direct way.[14] Evidently, men are trained to act out their anger, not to reflect on it. Therefore, they may hit objects, raise their voices during arguments, and use unfiltered language. Likewise, the male mind is less likely to recognize anxiety; it tends to express it as anger rather than concern.[15] For this reason, in the face of anxiety, a man may display constant tension or feel compelled to control things and people. On the other hand, the female mind generally has less difficulty admitting anxiety and even seeking professional help when necessary.

When a woman is emotionally distressed, what does she do? She talks, cries, seeks support, asks for help, and does not usually calm down until she finds it. And what does a man typically do? The first thing to note is that when a man is emotionally overwhelmed, he rarely asks for help. Instead, he prefers to retreat into his cave. His distress disconnects him; he appears silent, isolated, with poor eye contact. He tends to avoid situations and seeks refuge in work or tasks that keep him immersed in his own world.

When his amygdala becomes paralyzed, he may act with coldness and indifference, as if nothing matters. His mind seems to go blank. He does not respond during discussions. When fear places him in a state of hypervigilance, he feels constantly threatened. As a result, he becomes defensive, interprets neutral comments as personal attacks,[16] and lives as if he were about to explode.[17] He does not directly say, "I am afraid of...," but he withdraws because of... That is precisely why he avoids difficult conversations, changes the subject, maintains distance to "relax," or walks alone without knowing where he is going.[18]

When the threat targets his self-esteem, a man may experience shame and guilt, which later transforms into aggression.[19] This leads to self-criticism, emotional self-punishment, and deep silence.[20]

How Do the Amygdala and the Prefrontal Cortex Work Together?

The way the amygdala and the prefrontal cortex align to work together is truly interesting. When the amygdala detects the presence of danger, it immediately sends information to the brain to activate intense emotions. This signal is received by the prefrontal cortex, which evaluates the content of that information. It then considers whether it is a real threat or something irrelevant, in order to determine whether it should actually act. However, that is not all. The prefrontal cortex goes a step further to confirm whether the action can be carried out within a safe framework and analyzes whether it is appropriate to respond in that way.

What happens then? The prefrontal cortex, if it deems it necessary, regulates the response. It may choose to calm the amygdala, reduce impulses, or even guide toward a more intelligent response. At that point, a type of dispute arises between emotion and reason.

Is it possible for the amygdala to "hijack the mind" and, instead of generating a rational response, impose an emotional one? According to numerous scientific studies,[21] the answer is yes.[22] The concept of "mind hijacking" or "amygdala hijack" should be understood as a figurative term. When an emotion such as anger, panic, anxiety, or another reaches high intensity, it can block the prefrontal cortex and prevent it from performing its regulatory function. The term "hijack the mind"[23] was initially used by Daniel Goleman,[24] based on earlier studies conducted by Joseph LeDoux and his colleagues.[25]

In the face of provocation, the amygdala can activate in milliseconds; in contrast, the prefrontal cortex typically requires more time. Popularly, some suggest that it takes between 6 and 8 seconds to activate, although scientifically it is difficult to specify an exact interval. The time required for the prefrontal cortex to process and regulate information will largely depend on cognitive complexity and the nature of the stimulus.

If, in a critical moment, a person loses control and reacts with anger without thinking about what they say or do—if their words are hurtful and their decisions appear erratic or imprudent—this indicates that the amygdala has won the battle over the brain. Some might say: "The provocation was so great that it was impossible to control oneself, and reacting that way was inevitable. As you can see, I had nothing to do with it." Would this be a prudent line of reasoning, or an excuse intended to justify a lack of self-control?

When the prefrontal cortex is active and well-trained, in situations like these it is the structure that prevails. As a result, what we truly need to see occurs: behavior can be regulated by reason rather than governed by emotion. Anger does not dominate but is controlled; consequences are evaluated, and the person can act with maturity, choosing strategic rather than impulsive responses. It is in these circumstances that empathy, morality, and the pursuit of the common good can emerge.

When the amygdala feels but allows the prefrontal cortex to make decisions, the desired behavioral balance is achieved. This is something possible for every man who desires it.

Let Us Consult the Bible and Complete the Idea...

1. Proverbs 14:29 says: "He who is slow to anger is...

   ____

2. On the other hand, Proverbs 29:11 emphasizes that "a fool... ______________________________
   ____."

3. On the other hand, Proverbs 29:11 emphasizes that "a fool... ______________________________
   ____."

4. According to Proverbs 15:18, there is one thing that an angry man certainly does—what is it? __________
   ____.

   But "he who is slow to anger... ____________________
   ____."

5. For a successful masculinity, self-control and emotional regulation are more valuable than physical strength. How is this concept summarized in Proverbs 16:32?

   ____

   ____.

6. How does Proverbs 25:28 describe the man whose spirit has no restraint? What does that mean to you?

   ____

   ____

How to Train the Brain to Control Anger

The ability to demonstrate self-control in challenging circumstances clearly reveals that the mind has been trained and has

developed the strength necessary to achieve a healthy balance.[26] Therefore, training the male brain for effective anger control is a possible mission.

How is this achieved? This goal is not reached through force, but through a process that involves emotional and spiritual reeducation, strengthening self-control, and creating new neural pathways that replace aggressive impulses with strategic and appropriate responses. Clearly, this requires an integrated effort involving the different dimensions of the human being—physical, mental, emotional, and spiritual—so that results are reflected in behavior. Let us consider some practical recommendations:

1. Train Your Body Through a Physiological Pause

Can we expect that at some point you will be seriously provoked? We may not know when or how it will happen, but most likely it will. At that moment, the first thing you must do is apply the brakes to your body's emotional discharge. Take note: the goal here is not to think—it is to stop. To do this, I recommend these simple yet very effective techniques:

a. **4–2–6 Breathing Method:** Inhale for 4 seconds, hold for 2 seconds, and exhale for 6 seconds (12 seconds total). The prefrontal cortex takes longer to activate than the amygdala. This exercise gives you the opportunity to gain better control of the situation. For optimal effect, it should be practiced for at least 10 minutes, reducing negative emotion and creating a more relaxed emotional state.

b. **Distance yourself from the trigger for at least 90 seconds.** This is usually enough time to significantly reduce the chemical discharge associated with anger. After that, you will likely feel better equipped to handle the situation.

c. **Relax.** Your jaw may be tense—release it, and do the same with your shoulders and hands.

d. **Lower your voice deliberately.** No matter how strongly you feel like raising it, make a conscious effort to speak more softly. Your body will tend to follow the tone of your voice.

Practice training your body in this way. As you develop this habit, you will reduce those costly impulsive reactions. It may be difficult at first, but do not worry—that is part of the process. Great changes are the result of small improvements practiced consistently.[27]

2. P–R Training: Think First, Then React

When anger is activated, the brain quickly constructs emotionally charged narratives:

"They are abusing me..."

"They attacked me..."

"That is unacceptable..."

"They think they can control me..."

What resources can you use to counter this tendency?

a. **Pause.** Remember that haste is the mother of many mistakes.

b. **Ask yourself honestly:** With the information I have, what am I interpreting? Then ask: Is this interpretation 100% true, or is it driven by emotion?

c. **Consider other perspectives.** If others had the same information, what explanations might they offer?

This training strengthens prefrontal pathways, promoting more conscious, mature, and balanced responses.

3. Evaluate Your Frustration Tolerance

As men, we often enjoy feeling in control. However, the frustration that arises when we lack control must be learned and managed.

Examples:

a. If you arrive at a place where you have to stand in a line and you are not among the first, remaining in it without complaining or ruminating negative thoughts will help you develop greater patience.

b. In the city where you live, at specific hours, sometimes frustrating traffic forms. Understanding that you will not always have control will allow you to enter into context and work better on frustration.

c. If you have access to gratification and you are willing to postpone it; you will be doing yourself a good of extraordinary proportions.

These practices reduce explosiveness and improve emotional regulation.

4. Train Yourself to Express Emotions in a Healthy Way

Many men were trained to suppress emotions. Over time, this repression turns into anger. What can you do?

a. Express what you feel without attacking.
b. Make your statements in the first person: use “I,” instead of saying “you always...” or “you never...”.
c. An example would be: “I felt ignored; I would like to be taken into account...”.

This reduces defensiveness and prevents escalation.

5. Care for Your Body

A disciplined body supports emotional regulation:

a. Take time to exercise regularly. You may need to reorganize your priorities, but this practice will help you reduce cortisol and improve impulse control.
b. Try to sleep between 7 and 8 hours daily. This habit contributes to the recovery of the prefrontal cortex and keeps it in a better operational condition.
c. If you adopt a diet low in sugar, you will improve your physical and emotional health. If you also control fat and salt, the benefits will be even greater.

Good physical health improves emotional control.

6. Redefine “Strong Man”

Many were taught that strength equals dominance or toughness.

a. What if, instead of defining masculine strength as the ability to shout, dominate, and impose oneself over anyone, we review and elevate that concept?
b. The truly strong man is not the one who controls more people or things, but the one who controls himself. He is the one who can be firm and, even so, remain calm; the one who can speak clearly and express what he thinks and feels.

When we come to understand that the strongest man in the world is not the one who defeats his opponents, but the one who can defeat himself, that day we will have reached the greatest of victories.

My personal exercises

Exercise 1: Personal anger radar. This exercise will help me identify patterns and triggering factors.

a. Draw a circle divided into six parts.

b. In each section, detail:

- Behaviors of: a friend, my partner, my children, my boss, etc. that trigger my anger:
- Situations that frustrate me:
- Bodily stimuli that appear when I feel angry:
- Thoughts that quickly arise in my mind in those moments:
- Generally, these are my reactions:
- Consequences of those reactions (include the areas: emotional, family and social relationships, spiritual life):

____

____

____

____

____

Exercise 2. Improving my response / reaction

This exercise will help me change reactive impulses into reflective decisions.

Think about a recent event in which you became seriously angry and now describe:

a. What was it that happened?

____

b. When that happened, what did I think?

____

c. Then, this is what I felt...

____

d. And I reacted in this way...

____

e. But, being honest, I think it would have been better if I had done this:

____

Exercise 3. My action plan for difficult moments

I am aware that critical events may arise at any moment; therefore, from now on I will prepare a clear path for such cases:

a. When I feel angry, I will pause for ______ seconds.

b. My control phrase will be: (some people prefer to have a motivational phrase, a biblical verse such as Phil. 4:13, etc.).

c. If I see that I need space, this is what I will say: ______________________________.

d. If I feel the need to express my feeling of discomfort, I will do it with respect and self-control, using phrases such as:

"When ______________ occurred, I felt ______________. I would like that next time we can ______________."

From man to man...

Building a strong man

Resources: Cards, flip charts, and markers.

From man to man, let us try to redefine a masculinity far from anger:

a. On one side of the card, write what it means to be a "strong man" according to what you learned at home, from your friends, or from your culture.

b. On another card, define a new model of masculine strength that includes characteristics such as:

- Controlling impulses
- Speaking calmly
- Listening attentively
- Respecting others
- Having a willingness to forgive

My Personal Reflections

____

____

____

____

____

Endnotes:

1. De León, C., & De León, C. (2024). *Construyendo relaciones sanas.* Pacific Press.

2. Menon V, D'Esposito M. The role of PFC networks in cognitive control and executive function. Neuropsychopharmacology. 2022 Jan;47(1):90-103. doi: 10.1038/s41386-021-01152-w. Epub 2021 Aug 18. PMID: 34408276; PMCID: PMC8616903.

3. Harrewijn A, Ruiz SG, Abend R, Haller SP, Subar AR, Swetlitz C, Valadez EA, Brotman MA, Chen G, Chronis-Tuscano A, Leibenluft E, Bar-Haim Y, Fox NA, Pine DS. Development of Neural Mechanisms Underlying Threat Processing: Associations With Childhood Social Reticence and Adolescent Anxiety. Biol Psychiatry Glob Open Sci. 2023 Feb 3;3(4):893-901. doi: 10.1016/j.bpsgos.2023.01.008. PMID: 37881548; PMCID: PMC10593903.

4. Baumeister, R. F., & Vohs, K. D. (2016). Executive function, the brain, and self-regulation. In K. D. Vohs & R. F. Baumeister (Eds.), *Handbook of self-regulation: Research, theory, and applications* (3rd ed., pp. 3–25). Guilford Press.

5. Greene JD, Nystrom LE, Engell AD, Darley JM, Cohen JD. The neural bases of cognitive conflict and control in moral judgment. Neuron. 2004 Oct 14;44(2):389-400. doi: 10.1016/j.neuron.2004.09.027. PMID: 15473975.

6. Bechara A, Damasio H, Damasio AR. Emotion, decision making and the orbitofrontal cortex. Cereb Cortex. 2000 Mar;10(3):295-307. doi: 10.1093/cercor/10.3.295. PMID: 10731224.

7. Ochsner KN, Ray RD, Cooper JC, Robertson ER, Chopra S, Gabrieli JD, Gross JJ. For better or for worse: neural systems supporting the cognitive down- and up-regulation of negative emotion. Neuroimage. 2004 Oct;23(2):483-99. doi: 10.1016/j.neuroimage.2004.06.030. PMID: 15488398.

8. Shamay-Tsoory SG. The neural bases for empathy. Neuroscientist. 2011 Feb;17(1):18-24. doi: 10.1177/1073858410379268. Epub 2010 Nov 11. PMID: 21071616.

9. Newberg, A., & d'Aquili, E. (2001). *Why God won't go away: Brain science and the biology of belief.* Ballantine Books.

10. Miller EK. The prefrontal cortex and cognitive control. Nat Rev Neurosci. 2000 Oct;1(1):59-65. doi: 10.1038/35036228. PMID: 11252769.

11. For more information on this point, consult Chapter 1 of the book *Spark Up Your Marriage,* by the same author.

12. Archer J. Does sexual selection explain human sex differences in aggression? *Behavioral and Brain Sciences*. 2009;32(3-4):249-266. doi:10.1017/S0140525X09990951

13. Kimmel, M. (2017). *Manhood in America: A cultural history* (4th ed.). Oxford University Press.

14. Archer, J. (2004). Sex Differences in Aggression in Real-World Settings: A Meta-Analytic Review. *Review of General Psychology, 8*(4), 291–322. https://doi.org/10.1037/1089-2680.8.4.291

15. American Psychological Association. (2019). *APA guidelines for psychological practice with boys and men*. Washington, DC: APA.

16. Beck, J. S. (2011). *Cognitive behavior therapy: Basics and beyond* (2nd ed.). Guilford Press.

17. Gross, J. J. (2014). Emotion regulation: Conceptual and empirical foundations. In J. J. Gross (Ed.), *Handbook of emotion regulation* (2nd ed.). Guilford Press.

18. Real, T. (1998). *I don't want to talk about it: Overcoming the secret legacy of male depression*. Scribner.

19. Gilligan, J. (2003). *Violence: Reflections on a national epidemic*. Vintage.

20. Brannon, R. (2015). The male role and emotional suppression. In R. F. Levant & Y. Wong (Eds.), *The psychology of men and masculinities*. APA Press.

21. Arnsten AF. Stress signalling pathways that impair prefrontal cortex structure and function. Nat Rev Neurosci. 2009 Jun;10(6):410-22. doi: 10.1038/nrn2648. PMID: 19455173; PMCID: PMC2907136.

22. Hur J, Smith JF, DeYoung KA, Anderson AS, Kuang J, Kim HC, Tillman RM, Kuhn M, Fox AS, Shackman AJ. Anxiety and the Neurobiology of Temporally Uncertain Threat Anticipation. J Neurosci. 2020 Oct 7;40(41):7949-7964. doi: 10.1523/JNEUROSCI.0704-20.2020. Epub 2020 Sep 21. PMID: 32958570; PMCID: PMC7548695.

23. Goleman, D. (1996). *Emotional Intelligence*. Bantam Books.

24. Goleman, D. (2006). *Inteligencia social: La nueva ciencia de las relaciones humanas*. Editorial Kairós.

25. LeDoux, J. E. (1996). *The emotional brain*. Simon & Schuster.

26. White, E. G. (2009). *La educación* (ed. revisada). Asociación Casa Editora Sudamericana.

27. Clear, J. (2018). *Atomic habits: An easy & proven way to build good habits & break bad ones*. Avery.

CHAPTER 4

TO SPEAK OR NOT TO SPEAK

A woman sought counsel seeking help to resolve a problem that, in her view, was quite serious. Her husband hardly spoke, and she honestly acknowledged that her own style was completely the opposite.

"Please, help me," she said with evident concern. "I need to fix this man."

The counselor looked at her calmly and, among other things, asked: "Have you noticed any recent changes in his behavior?"

"No," she replied. "This is not something new. In fact, he has always been this way. The problem is that he doesn't communicate. He doesn't talk, he doesn't express himself...

and I have the feeling that we live in two completely different worlds."

She paused for a moment and added: "I need to talk in order to understand, to connect. He, on the other hand, remains silent, as if that were enough. And that silence, for me, feels like distance."

The Male Brain and Communication

Within the context of this discussion, a question naturally arises: is the male brain designed to communicate less?

Various speakers have popularized the idea that there is a significant difference in the number of words spoken by men and women. It has often been repeated—through data, anecdotes, or simple observations—that women speak around 20,000 words per day,[1] while men[2] speak only 7,000.[3] These figures became widely accepted.[4] However, to date, this information has not been scientifically validated.

The topic, however, has continued to generate interest. A key study that provided scientific evidence was conducted by the University of Arizona and led by Matthias Mehl and his colleagues. The study, titled "Are Women Really More Talkative Than Men?", included 396 participants (210 women and 186 men), mostly university students. Over several days, multiple audio recordings were collected. The findings indicated that there is no significant difference between women and men in the number of words spoken daily. Women averaged 16,215 words per day, and men 15,669. The research was published in July 2007 in Science, one of the most prestigious scientific journals in the world, and was later cited in PubMed[5] and other academic sources.

When analyzing this study, some people pointed out that the sample size was relatively small and limited to a specific age range, and therefore considered the data insufficient to establish definitive conclusions. However, in February 2025, a larger-scale study was published, including 2,197 participants (adolescents, young adults, and adults) and analyzing more than 631,000 audio recordings. The results showed that men spoke an average of 11,950 words per day, while women spoke 13,349.[6] Do this data suggest a statistically significant difference? The findings do not appear to support that conclusion.

A more precise question then arises: what is the real communicative difference observed between men and women? It is evident that the apparent scarcity of male communication, from a biological standpoint, is not due to an inherent limitation of gender. Communication is always taking place, even in the absence of words. Therefore, this limitation must be sought primarily in exogenous factors—cultural, social, and learned—that shape the way many men have been trained to express (or suppress) their inner world. Let us consider some of these factors:

How Men Tend to Communicate

Men tend to be direct and focus on the central point of their concern, with the intention of finding a solution. Women, on the other hand, tend to be more emotional and expressive. While men seek to solve the problem, women seek to understand the reason behind the problem. Male thinking is oriented toward solutions; female thinking toward connection.

In many men, there appears to be a limited and learned ability to identify, name, and manage emotions. It is important to clarify that when we speak of alexithymia,[7] we are not referring to a disease in itself,[8] but rather to a trait that may help explain why some men struggle to find words to express their emotional world.[9] The important point is that this trait can be effectively addressed through proper training.[10]

Cultural Norms

In certain contexts, it is not considered appropriate to see a man displaying sensitivity or tenderness. If he does, it may become a source of surprise for other men and even for some women.

Expressing need and reaching the point of asking for help is perceived, within this framework, as something incompatible with "true manhood."

Family Pattern

Finally, there is the family pattern, a highly relevant factor in understanding this difference. The home is a child's first school[11] and the foundation of everything later expressed in society. It is there that one learns how to behave and how to respond to stimuli. The family models, shapes, and reinforces; society receives and validates, and that impact returns to the home, since it is also an open system.

In practical terms, a child who was raised with messages such as: "men don't cry," "endure in silence," "figure it out on your own," or "be careful not to show weakness," will likely experience difficulties in emotional communication—even if his linguistic capacity is fully intact.

The Value of Communication

According to Jack O. Balswick, communication is the heart of family life.[12] Being able to express our feelings and emotions appropriately is a true richness. What we say, how we say it, and especially the message conveyed through our nonverbal language are factors that determine the quality of our relationships and the kind of "atmosphere" that surrounds us.

There is no doubt that we all desire to have good relationships with others, especially with those closest to us. In fact, healthy relationships mark success in business, friendships, love, and even spiritual life. For example, John 17:3 defines and conditions eternal life as a relationship that is not only of knowledge but also of experience with God, highlighting the relational dimension of human beings and the need for effective communication to establish it.

For communication to function, it is vital to express what is on our minds (to communicate), to pay attention to what

is being said to us (to listen), and to grasp or understand the intended message (to assimilate). Unfortunately, the sender's intention is not always understood by the receiver. It is estimated that between 60% and 70% of senders believe they communicate clearly. However, studies by Thomas Gilovich suggest that only between 30% and 40% of receivers interpret the message as intended.[13] Can this generate serious conflict in any relationship? Absolutely.

The issue is not that people desire conflict, but that they have not yet learned to understand as we understand or to see as we see.[14] Therefore, our communication must create a flow of mutual understanding that makes strong relationships possible. It is a skill we are not born with, but one we can certainly cultivate daily.

Male communication will be more effective if we take care of the intention behind it:

Am I using it to focus on myself or to focus on you?[15]
In what way?
How can I use it for edification and emotional healing?

Let Us Consult the Bible and Complete the Idea:

1. James 1:19 – "So then, my beloved brethren, let every ____________ be ____________, slow to ____________, slow to anger."
2. According to Ephesians 4:15, the truth must be spoken. But it must always be spoken with ____________.
3. Ephesians 4:29 – "Let no ____________ corrupt word proceed out of your mouth, but what is good for ____________ ____________, that it may give grace to the hearers."
4. Colossians 4:6 – "Let your speech always be with ____________, seasoned with salt, that you may know how you ought to ____________ each one."
5. Proverbs 15:1 – "A ____________ answer turns away wrath, but a ____________ word stirs up anger."

Principles for a Fit Communication

An emotionally fit man uses communication strategically. Through it, he builds, clarifies, leads, restores, and protects those he loves. It is not something automatic, but a powerful tool for emotional growth that can and must be trained. To achieve this, let us consider some basic principles that characterize this model of masculinity:

1. Speak with Clarity and Purpose

The emotionally fit man does not speak merely to release emotions.[16] He speaks with a clearly defined purpose: to provide direction, support, and edification. He avoids ambiguity and is intentional in expressing what happened, what is needed, when it is needed, and what a possible solution might be in the face of disagreement.

Listen Before Speaking

Active listening improves the quality of interaction between people. When someone feels heard, their performance improves, both in the workplace and in the family.[17] In many relationships, the problem is not bad intention, but the absence of active listening, which ultimately affects communication.

Practice Empathy and Emotional Validation

Even when he does not agree with everything, he knows how to put himself in the other person's place in order to understand their inner world. When empathy is present, relationships flourish, satisfaction increases, and both emotional pain and anxiety levels decrease.[18]

Live Assertiveness

When he has needs, he expresses them; when he requires boundaries, he establishes them. He understands the balance between firmness and respect. If things are not going as he desires, he speaks clearly, in the first person, and trains himself for increasingly healthy communication.

Maintain Verbal and Nonverbal Consistency

When he speaks—and also when others speak to him—the fit man is present both physically and mentally. His gaze, facial expression, tone of voice, body posture, and the content of his words communicate the same message: "I am here for you."

Nonverbal signals, in particular, convey very powerful relational messages.[19]

Seek Feedback and Establish Agreements

He is not afraid to ask for feedback regarding the behavior others observe in him. He understands that this information can help him grow. From there, both parties can reach clear agreements about what to do differently and in which areas change will occur.

Practice Repair

This man understands a fundamental principle: "If you do not want to walk through life broken, you must learn to repair." When tensions arise, he seeks to prevent problems from escalating: he respects speaking turns, takes pauses, intentionally pursues reconciliation, acknowledges his part in the conflict, knows how to apologize, and, when appropriate, uses healthy humor to reduce intensity. Where there is a spirit of repair, the likelihood of restoring a relationship increases significantly.

From Man to Man

As men, we speak, and among men we understand each other. Now allow me to ask you three critical questions, and I invite you to answer them with an open mind and a sincere heart:

1. When I speak to the people I love, am I communicating from the identity of the man I want to become, or from the wounded man who has not yet healed?
2. When I face conflicts at home, which pattern repeats most frequently: attacking, controlling, remaining

silent, or withdrawing? What is that pattern costing me on a personal, family, and spiritual level?

3. If within one year my relationships were a direct reflection of my current communication habits, would I feel proud of those outcomes? And if not, what specific habit must I begin to train today?

How to Train the Male Brain for "Fit" Communication

There are numerous resources that can be very helpful for those who wish to improve those common tendencies that sometimes make us appear distant or disconnected from those we love most. As men, we know what commitment means and how important it is for achieving the adjustments that make us better individuals, better partners,[20] and better friends. This is precisely one of the first steps we must take. Let us consider some of them:

Exercise 1: Training Emotional Language

This week, take between 5 and 10 minutes each day to:

- **My 3-word check-in:** Try to do it at the same time each day and use three words to describe how you feel.

Example: "Today I feel ________, ________, ________."

- **A complete sentence**: Label or express a feeling, its cause, and your need.

Example: "I feel ________ because ________, and I need ________."

This training strengthens the ability to recognize emotions and develop less defensive responses.

Exercise 2: Connect Instead of Solve – The 20/80 Principle

Before offering verbal or practical solutions, focus first on asking these two questions:

- "Do you want me to listen or help solve it?"
- "What would support look like for you right now?"

In many cases, a woman does not need the man to solve the problem, but rather to provide emotional connection and attentive listening.

Exercise 3: The "Reflection" Technique – Key to Active Listening

Especially useful in difficult conversations:

- Paraphrase their words: "What I hear you saying is..."
- Validate the feeling: "I understand... it makes sense that you feel this way..."
- Ask: "What part is the most difficult for you?"

This technique helps reduce tension and promotes emotional connection in the relationship.

Exercise 4: Weekly Family Ritual

Set aside 20–30 minutes without screens (no TV, phone, or computer).

Basic rules:

- Each person has equal time to speak (1–3 minutes, as agreed).
- No interruptions.
- No sarcasm or disrespect.
- Strict confidentiality.
- If a major issue arises, schedule it for another time.

Suggested format:

Gratitude (1 minute each):
What are you grateful for?

Calendar (5–8 minutes):

- Weekly schedule (work, school, appointments, commitments).
- Responsibilities and needed support.
- Family finances (payments, expected expenses).
- "Defuse the grenade": anticipate possible tensions and agree on strategies.

Emotional Topic (3–7 minutes):
How did I feel this week?

- Something that made me happy / something that weighed on me.
- A current emotional need.
- Area of personal growth and emotional goal for the week.

Special closing:

This exercise can be adapted for couples or family/social relationships. Many couples choose to close with a prayer, asking for God's blessing on what was shared.

This training, when practiced consistently, improves communication and strengthens marital and family empathy. It also helps expand the couple's "love map"[21] and address accumulated conflicts with greater clarity, promoting emotional security.

If conflict escalates, it is human for emotions to intensify. This does not mean the relationship is destined to end. If irritation increases during dialogue, it may be healthy to take a pause and allow the prefrontal cortex to regain control. A break of about 20 minutes can help regulate emotions. During that time, one may walk, practice breathing, drink water, and return to calm before resuming the conversation.

My Personal Reflections

____

____

____

____

____

Endnotes:

1. Brizendine, L. (2006). *The female brain*. Broadway Books.
2. It should be noted that the cited reference was later revised to add a more investigative and scientific tone.
3. Liberman, M. (2006, septiembre 24). *Sex on the brain*. University of Vermont, Science for the People Listserv Archives. https://list.uvm.edu/cgi-bin/wa?A3=ind0609&B=----------------------------1159136603&E=quoted-printable&L=SCIENCE-FOR-THE-PEOPLE&P=1171728&T=text%2Fhtml%3B+charset%3DISO-8859-1

4. Figure popularized in the 1970s–80s and later in the 2000s. The latter by speaker Mark Gungor, who cited in his presentations that men needed to speak 10,000 words per day, while women needed 20,000. The full video can be found at: Amara. (February, 2021). *Tale of Two Brains full* [Video]. Amara. https://amara.org/videos/pMFT7NeTwu77/es-mx/1140514/.com

5. Mehl MR, Vazire S, Ramírez-Esparza N, Slatcher RB, Pennebaker JW. Are women really more talkative than men? Science. 2007 Jul 6;317(5834):82. doi: 10.1126/science.1139940. PMID: 17615349.

6. Tidwell CA, Danvers AF, Pfeifer VA, Abel DB, Alisic E, Beer A, Bierstetel SJ, Bollich-Ziegler KL, Bruni M, Calabrese WR, Chiarello C, Demiray B, Dimidjian S, Fingerman KL, Haas M, Kaplan DM, Kim YK, Knezevic G, Lazarevic LB, Luo M, Macbeth A, Manson JH, Mascaro JS, Metcalf C, Minor KS, Moseley S, Polsinelli AJ, Raison CL, Rilling JK, Robbins ML, Sbarra D, Slatcher RB, Sun J, Vasileva M, Vazire S, Mehl MR. Are women really (not) more talkative than men? A registered report of binary gender similarities/ differences in daily word use. J Pers Soc Psychol. 2025 Feb;128(2):367-391. doi: 10.1037/pspp0000534. Epub 2025 Jan 13. PMID: 39804347; PMCID: PMC11825285.

7. Alexithymia is known as the difficulty that some individuals present in identifying and expressing emotions. This does not mean that they lack them, but rather that they are not able to verbalize them. Emotional distancing, cold communication, and the lack of tools to empathize often generate significant conflicts with their partner, as they are perceived as voluntary coldness or lack of interest in the relationship. Among other things, it is primarily a problem of early attachment styles, as well as adaptive responses, as in the case of secondary alexithymia. In the latter are included conditions such as depression, anxiety, chronic stress, traumatic events, or even the practice of discouraging the person from expressing their emotions when it was understood to be appropriate to express them. For more information on this point click the link below:

8. Speranza, A., De & Todarello, O. (Eds.). (2020). *Alexithymia: Advances in research, clinical practice, and neurobiology*. Cambridge University Press. https://www.amazon.com/Alexithymia-Advances-Research-Clinical-Practice/dp/1108416640

9. Levant RF, Parent MC. The development and evaluation of a brief form of the Normative Male Alexithymia Scale (NMAS-BF). J Couns Psychol. 2019 Mar;66(2):224-233. doi: 10.1037/cou0000312. Epub 2018 Oct 4. PMID: 30284847; PMCID: PMC10388695.

10. Tsubaki, K., & Shimizu, E. (2024). Psychological Treatments for Alexithymia: A Systematic Review. *Behavioral Sciences, 14*(12), 1173. https://doi.org/10.3390/bs14121173

11. White, E. G. (2009). *La educación*. Asociación Casa Editora Sudamericana.

12. Balswick, J. O., & Balswick, J. K. (1999). *The family: A Christian perspective on the contemporary home* (2nd ed.). Baker Academic.

13. Gilovich, T., Savitsky, K., & Medvec, V. H. (1998). *The illusion of transparency: Biased assessments of others' ability to read one's emotional states. Journal of Personality and Social Psychology, 75*(2), 332–346. https://doi.org/10.1037/0022-3514.75.2.332

14. Covey, S. R. (2005). *Los 7 hábitos de las familias altamente efectivas*. Paidós.

15. Tripp, P. D. (2017). *Guerra de palabras: Tratando el corazón de tus problemas con la comunicación*. Pioema Publicaciones.

16. Covey, S. R. (1999). *Los 7 hábitos de las familias altamente efectivas* (The 7 Habits of Highly Effective Families). Editorial Paidós.

17. Weger, H., Castle Bell, G., Minei, E. M., & Robinson, M. C. (2014). The Relative Effectiveness of Active Listening in Initial Interactions. *International Journal of Listening, 28*(1), 13–31. https://doi.org/10.1080/10904018.2013.813234

18. Howcroft, A., Bennett-Weston, A., Khan, A., Griffiths, J., Gay, S., & Howick, J. (2025). *AI chatbots versus human healthcare professionals: A systematic review and meta-analysis of empathy in patient care*. British Medical Bulletin, 156(1), Article ldaf017. https://doi.org/10.1093/bmb/ldaf017

19. Patterson, M. (2018, June 25). Nonverbal Interpersonal Communication. *Oxford Research Encyclopedia of Communication*. Retrieved 6 Feb. 2026, from https://oxfordre.com/communication/view/10.1093/acrefore/9780190228613.001.0001/acrefore-9780190228613-e-657.

20. Hendrix, H. (2019). *Getting the love you want: A guide for couples*. St. Martin's Griffin.

21. Gottman, J. M., & Silver, N. (2015). *The seven principles for making marriage work: A practical guide from the country's foremost relationship expert* (Revised ed.). Harmony Books.

CHAPTER 5

THE MAN AND HIS SEXUALITY

From a young age, Mark learned that sexuality was something to be demonstrated, not something to be understood. For him, desire was impulse, accumulated tension, a need for release. No one ever taught him that his sexuality was also connected to his identity, his ability to bond, and his emotional world. He grew up believing that being a man meant performing, responding, not failing. Thus, when he fell in love with Christy, he assumed that loving meant fulfilling, and that if there were problems, silence was the best defense.

Christy, on the other hand, experienced sexuality as an extension of emotional connection. For her, feeling heard, valued, and safe was an essential part of intimacy. Mark did not understand this. He became frustrated when she asked for conversation instead of physical closeness, and she felt used when he sought closeness without dialogue. They both loved each other, but they spoke different languages. He did not know how to name what he felt; she did not know how to ask without hurting. Confusion grew, and with it, the distance widened.

Mark began to feel less like a man. He did not understand why his desire was so strong and constant, while his ability to express emotions was so limited. He blamed himself, shut down, and avoided difficult conversations. Inside, there was shame, fear of not being enough, and a fragile identity sustained only by his role and performance. Christy, tired of trying alone, began to doubt the future of the relationship. Many thoughts crossed her mind...

The turning point came when, in the midst of a crisis, they agreed to seek help. They learned that their differences were not defects, but distinct designs that required understanding and adjustment. Mark began to train his ability to speak about emotions and to listen without becoming defensive; Christy learned to express her needs without accusing. They discovered practical solutions: scheduled times for conversation, clear agreements, respect for each other's rhythms, and mutual boundaries.

Faith became an anchor. They prayed together—not to change one another, but to be transformed themselves by the extraordinary power of their Creator. Mark learned that his worth did not depend solely on his sexual performance, but on his character and his ability to love. Christy found security in seeing a man willing to grow. Prayer and a growing relationship with Christ gave them patience, humility, and hope.

Today, their relationship is stable—not because they stopped being different, but because they learned to walk together. Mark understood that his sexuality is part of his identity, but not its entirety; Christy understood that growth is a process. Together, they discovered that when there is a clear identity, honest dialogue, and a living faith, intimacy ceases to be a battlefield and becomes a place of connection.

Male Sexuality

Human sexuality is an integral part of every individual. It accompanies us throughout life and is composed of a variety of biological, psychological, emotional, social, cultural, and spiritual factors. Genesis 1:28 declares: "And God blessed them, and God said to them, 'Be fruitful and multiply; replenish the earth...'" Later (2:24), sexuality is presented as part of God's plan for marriage. Clearly, sex did not involve only the physical. God created man and woman, and He made each with biological and emotional characteristics that distinguish them. Rather than allowing these differences to generate tension, there is also the possibility of appreciating them, enjoying them, and celebrating the beauty of being different.

What does sexuality mean for a man? The story of Mark and Christy offers us a window into the two worlds that often characterize male and female mindsets regarding sexuality. For him, being successful in the bedroom holds critical importance. Very often, this functions as an indicator of his self-worth, how competent he perceives himself to be, and as a silent confirmation of whether he is truly a man or not. If this "thermometer" of masculinity is not functioning well—if he feels rejected by his partner, if there is any dysfunction, or if he is compared—the male mind does not interpret it as an erotic situation that needs to be worked on, but as a direct blow to his identity. Generally, he does not talk about it; he keeps it within.

In some cases, the male mind views sex as the ideal means to reduce tension, as an emotional outlet, and as a way to regulate stress and anxiety. Seen from this angle, it is as if the body becomes the "mouth" of a man who has had difficulty articulating his emotions or who never learned how to do so. Likewise, influenced by a distorted view of the male role, some come to see women as beings who exist to serve men, with the assumption that he possesses authority, power, and dominance over her. If he succeeds in keeping her "submissive," then he perceives himself—and is valued by others—as a man who has established his authority in the home.

If, by any chance, a woman is reading these lines, she might think: "Oh, if this is the case, then man is a monster!" It is fair to clarify and say: NO. Man is not a monster. He is simply different from woman. Moreover, what has been described above does not represent the full picture of the male mindset regarding sexuality.

There is another lens through which male sexuality can be viewed. In it converge values, ethics, and the spiritual component. This lens refuses to see sexuality as a matter of dominance, power, or authority. Far from being merely an impulse, it sees it as an expression of a higher purpose that goes beyond temporary desire. For this man, sexuality is the vehicle through which the body expresses the living love that the heart already feels. What for other men is internal struggle or repression becomes, for him, an exercise of self-control and an opportunity for constant growth.

He learned this in the school of Christ, and following the example of his Model, he chooses to remain faithful to his marital covenant. He cares for his partner, shows her trust and deep love, stands by her in good times and in bad, is intentional about cultivating a strong emotional connection, and in moments of crisis, is willing to be part of the solution. The integrative lens views sexuality and immediately recognizes it as a gift from God to humanity: to be enjoyed, for the multiplication of the species, and as an illustration of the quality of spiritual connection between the Creator and His creatures.

The Biological Component

It is necessary to emphasize that male sexuality must be viewed from an integral perspective and not merely as some-thing

genital or physical. Beyond the body, it involves the whole being: the mind, emotions, the quality of relationships, and the spiritual component. For example, if a man has his genital organ in good condition but carries unresolved issues such as guilt, shame, low self-esteem, identity problems, or trauma, his sexual response may be seriously affected.

With regard to the biological aspect of male sexuality, in a previous chapter—when addressing the topic of sexual differences between men and women—we explained that the XY sex chromosome determines male biological sex. There we observed that testosterone production directly influences sexual desire, impulsivity,[1] energy, and competitive behavior. The greater the amount of properly functioning testosterone, the greater the overall activity in terms of semen and sperm production. In men, this process is not cyclical but continuous.

What happens to the semen that is produced? The production that is generated accumulates. Where? Primarily in the seminal vesicles, the epididymis, and the vas deferens. Under normal conditions, this activity occurs constantly.[2] Based on various studies, during abstinence, the seminal vesicles increase significantly[3] within approximately 2 to 3 days,[4] reaching an optimal level of content.[5]

It is commonly said that overproduction creates an urgency for release. Could this experience be called an "urge"? To explain it simply, we could say yes. What happens in the male body? When the volume of seminal storage reaches these levels, the walls of the reservoirs stretch, and what we might call pressure receptors are activated. A signal is then sent to the sympathetic nervous system—something like an internal message that says: "We are in discharge mode."

For this reason, the man experiences a particular sensation in the pelvic area, sexual thoughts appear more frequently, and there is increased sensitivity to any erotic stimulus. From this perspective, the urge for release that a man feels is the result of a combined action of his mind, his brain, his hormones, and the mechanical response of his body. In other words, under normal conditions there is a direct connection between this high production and male sexual desire.[6]

Now, if stored production reaches its optimal level between 48 and 72 hours of abstinence, does this mean that ejaculation must necessarily occur? And if so, what harm could result if it does not? Let me explain it this way: when the biological condition is present, a discharge (ejaculation) occurs. Since production is continuous, after the mentioned time, the seminal vesicles reach that level again, and the natural behavior of the man once more suggests the need for another discharge.

If ejaculation does not occur, does production continue indefinitely? Here is where it becomes interesting: the way God designed the human body is extraordinary! Thanks to this design, the male body has a self-regulating system. If release does not occur through ejaculation, the body reabsorbs part of the seminal fluid or eliminates it through nocturnal emissions, also known as wet dreams.

Single, Young (and Handsome)

In contrast to the married man, who through his wife has a natural means of channeling his sexual desires, some single men feel they live in a serious tension of attraction. On one hand, their principles and values motivate them to be careful in managing their sexuality. On the other hand, when that single young man faces "urgent" desires to share with someone but does not have a wife with whom to be intimate, he asks himself: What do I do? Here is what some think and do to deal with this situation:

- **Casual Sex.** To decide what to do regarding their sexuality, some turn to their friends. As expected, in a postmodern and pluralistic society, many of those friends defend the idea that sex is free and life is short. For this reason, some end up engaging in sexual practices, whether occasional or frequent. It is, so to speak, quick sex without commitment. Certainly, life is short, but embracing the idea that sex is free tends to be more harmful than beneficial. In fact, it can turn a brief life into an even shorter, more complex, and less fulfilling experience. To take a quick look at a reality every man should consider—especially the single man—the Centers for Disease Control and Prevention reported that in 2024 alone, more than 2.2 million cases of chlamydia, gonorrhea,[7] and syphilis[8] were recorded in the United States. Not without reason, the apostle Paul tells us: "Flee sexual immorality..." (1 Corinthians 6:18).
- **Pornography**. A single young man once spoke with his mentor about his sexual needs. He described how he became trapped in pornography until he felt there was no way out. His time and best energies were invested there. He would simply tell himself: "I have no choice." This led

him to experience depression, anxiety, low self-esteem, difficulty connecting properly with others—especially with the opposite sex—and problems with his own sexual functioning.[9] Over time, this young man realized that what seemed like a harmless and pleasurable practice was not so. He also discovered that there were solutions to this problem. He found strength in God and professional help through a counselor who guided him along a path of restoration.

- **Social Media with Sexualized Content.** In this area, some engage in private messaging that may include suggestive images, sexual connotations, and constant visual stimulation. This practice can have a significant psychological impact. Among other things, it keeps sexual desire permanently activated, makes natural self-regulation more difficult, and promotes the concept of objectification.
- **Physical Activity.** Although intense physical exercise does not satisfy sexual impulses, it has a very positive effect. It reduces physiological activation, strengthens self-control, and promotes better management of immediate sexual gratification.[10] In other words, although sexual desire does not disappear, it has been shown that exercise can help reduce physiological urgency.[11] Activities that produce fatigue, competitive sports, and demanding physical routines help channel testosterone, aggression, and bodily tension, redirecting sexual energy.[12]
- **Spiritual Formation.** Engaging in spiritual disciplines has proven to be a positive factor in helping individuals regulate their emotions and respond more adaptively to life's realities. Important studies have shown that prayer, Bible reading,[13] meditation on its teachings, and trust in God's providence[14] facilitate the cognitive reframing of stressful situations,[15] promoting regulated rather than impulsive emotions.[16] In 2 Timothy 2:22, the Word of

God exhorts us: "Flee also youthful lusts, but pursue righteousness, faith, love, and peace with those who call on the Lord out of a pure heart." In a powerful way, the same apostle Paul affirms: "I can do all things through Christ who strengthens me" (Philippians 4:13).

"My Wife Wants Nothing..."

One of the most frequent complaints I hear from married men regarding their wives is the lack of harmony when it comes to sexual intimacy. "She is rarely willing...," "she never takes the initiative," or "those headaches never go away..." are some of the phrases often mentioned. If we interpret this literally, we might ask: is it that a woman's sexual desire is always inhibited? Or rather, is it that the way she operates is different?

Let us look at some relevant facts:

Men have more testosterone than women. One of the functions of testosterone is to increase spontaneous sexual desire. Since sexuality represents a source of gratification for men, under the influence of this hormone, they tend to become sexually stimulated more quickly[17] and actively seek this type of stimulation and pleasure.[18]

- It is often observed that men experience sexual desire as a physiological urgency that manifests in a more constant and direct way.
- For men, sexual desire does not necessarily depend on an immediate emotional bond. Even immediately after an argument with their partner, they may feel ready for sexual activity.
- For them, emotional bonding is strengthened during or after intimacy. For this reason, sexual rejection is often interpreted as personal rejection.

From a Sexual Perspective: Are Women "Strange"?

The answer to this question is broad. However, for reasons of space, we will limit ourselves to explaining some key points:

Do Women Think About Sex?

The idea that, from a sexual perspective, women are "strange" because their desire remains inhibited lacks scientific foundation. Male intensity and the apparent female "reserve" are not defects, but expressions of differences that, as men, we are called to explore and understand. It is not that women are not interested in sexual pleasure. In fact, some studies indicate that, on average, women think about sex about 10 times a day, compared to men, who average around 19.[19]

Let Us Separate Things

It is important to distinguish between low libido and a clinical disorder involving distress. Simply put, libido is the internal engine of sexual desire. In women, that desire tends to be more cyclical and less spontaneous. For example, the menstrual cycle (including premenstrual discomfort) and menopause are realities that men never experience, but women do. For this reason, some women report lower sexual desire during certain periods. However, when low libido is persistent and causes distress, anxiety, or impairment at a personal or relational level, we may be dealing with a disorder of female sexual interest/arousal, known as Female Sexual Interest/Arousal Disorder (FSIAD).[20]

Differences in Brain Activation

Studies in cognitive neuroscience have found that men exhibit greater activation in visual areas when observing suggestive or erotic stimuli in their partner.[21] Women, on the other hand, show greater activation in emotional and contextual networks. This helps explain why, in many women, sexual desire increases when the context of emotional intimacy is healthy. They integrate not only the amygdala and the prefrontal cortex, but also emotion, memory, and context. That is:

a. The quality of the bond that connects her to that man.

b. The experiences provided by the context (events of the day, positive experiences from yesterday, or negative ones that have not been properly resolved).

c. Based on daily interaction: What does that relationship mean to her?

d. What is her level of emotional security when she thinks about that relationship?

e. For her, emotional connection opens the door to sexual intimacy. When a woman gives sex without emotional connection, she generally feels used and, over time, shuts down.

All of this plays a fundamental role before speaking of a satisfying sexual experience. Sustained emotional connection—

more than compatibility or initial romantic intensity—is the strongest predictor of stability, satisfaction, and marital well-being.[22]

The question still remains: are women difficult? In reality, it is not that they are difficult; it is that they are different.

It is true that, in some cases, there is a disorder that makes a fuller expression of their sexuality more difficult. Some estimate that, depending on age range, the average lies between 6% and 12% of women.[23] Even so, these cases—which represent a minority—can generally be treated. What is interesting is that, for men, choosing to work on emotional connection with their partner could completely change the story.

Ten Principles for Strong Emotional Connection with Your Wife

1. **Show that you are truly emotionally available.**
   When she speaks to you, put your mind and emotions in "attention mode." Avoid distractions or giving the impression that you want to "fix things quickly."

2. **Listen with empathy.**
   The ability to listen to anything without interrupting, justifying yourself, or counterattacking is a virtue of men who operate at a higher level of emotional connection.

3. **Validate her feelings.**
   You may not agree with what she thinks, but still dare to recognize and value what she feels.

4. **Show genuine interest in her inner world.**
   Being a provider is important, but not sufficient. Discover where her thoughts are, what her fears are, what dreams she holds, and what struggles she faces today. That kind of conversation opens the door to her inner world.

5. **Communicate openly and honestly.**
   Do not be afraid to express what you feel. Whether it is

fear, pain, frustration, shame, or joy, do so with respect and maturity.

6. **Be sensitive to her emotional state.**
   Her emotional state is not always the same. When you notice changes, come alongside her, respond with care, and offer your support.
7. **Love her consistently, not conditionally.**
   Show verbal and physical affection without tying it only to sexual benefits. Be patient and persevering.
8. **Strengthen your relationship, not your pride.**
   The man who strengthens his relationship usually gains more than the one who only wins an argument. Winning her heart is worth more than imposing your reasoning.
9. **Build emotional security.**
   Create an environment where she feels safe to be

vulnerable. Be trustworthy, respectful, and emotionally stable.

10. **Strive to be an emotionally responsible man.** Acknowledge your mistakes, ask for forgiveness, and participate in healing the wounds caused. It will not always be easy, but if you ask for God's help, He will make it possible.

Let Us Consult the Bible and Complete the Idea:

1. "Therefore a man shall leave his __________ father and mother, and be joined to his __________, and they shall become __________________________." Genesis 2:24
2. "And God created man in His image... male and female He created them. Then God blessed them, and God said to them: _______________________ and __________________________..." Genesis 1:27–28
3. "Marriage is honorable among all, and the bed undefiled..." Hebrews 13:4
4. "Let the husband render to his wife the __________ __________, and likewise also the __________ to her husband..." 1 Corinthians 7:3–5
5. "Flee sexual immorality... your body is a __________________ of the __________ __________." 1 Corinthians 6:18–20

From Man to Man

There are three not-so-comfortable questions that every man must ask himself:

1. Does my sexuality build or destroy trust?
2. Am I using my partner, or am I loving my partner?
3. Am I being governed by my habits, or by the grace of God am I the one governing my habits?

Brain Training Exercises for Healthier Male Sexuality

The following exercises are designed to work on specific areas in men, such as impulse control, emotional regulation, bonding, and purpose.

Exercise 1: Conscious Delay of Impulse

This exercise strengthens self-control and the prefrontal cortex, allowing immediate sexual gratification to be postponed without suppressing desire.

What to do when sexual desire arises?

1. Stop for at least 90–120 seconds.
2. Take 4 to 6 deep breaths.
3. Ask yourself: Does this bring me closer to or further from the man I want to be?

After completing this exercise, decide what to do—but not during the peak of desire, rather after it has passed.

Impulsivity decreases when you allow the prefrontal cortex to intervene consciously. If the desire is not immediately reinforced, it tends to diminish. As a result, you will experience greater self-control and less impulsivity.

Exercise 2: Redirecting Sexual Energy

This exercise helps channel testosterone and redirect sexual energy toward productive and constructive actions.

1. When intense sexual desire is activated, what should you do?
2. For 15–30 minutes, engage in demanding physical activity such as weightlifting or competitive sports.
3. Afterward, breathe slowly for 5 minutes.

This exercise helps regulate testosterone, cortisol, and dopamine, reducing physiological urgency without denying desire. It is a practical way to lower bodily tension and improve emotional control.

Important note: Before starting this type of exercise, consult your doctor.

Exercise 3: Emotional–Verbal Training

This exercise integrates emotion with sexuality. Its purpose is to expand a man's repertoire of responses by regulating emotions and preventing sexuality from becoming the only available outlet for internal tension.

How to do it?

Practice it once a day. Identify and verbalize a real emotion:

- "I am frustrated..."
- "I feel anxious about..."
- "I feel alone..."

You may express it out loud or write it down. If you choose to write it, do not hesitate to expand or elaborate.

It is common for men not to verbalize emotions, but when they do, they reduce compensatory sexual behavior, strengthen emotional maturity, and develop a more conscious desire.

Exercise 4: Sexuality with Purpose

This exercise seeks to align sexuality with spiritual values—such as faith in Christ's power and a noble life purpose—instead of reducing it to mere impulses. It is designed to work on identity and desire.

How to do it?

- Once or twice a week, take time to reflect and write:
- What kind of man do I want to be sexually? Does my behavior honor God, my wife, and myself?
- Establish clear boundaries and motivations that inspire you to move closer to the ideal God has for you.
- Choose two Bible texts you would like to meditate on while doing this exercise.

When a man has a clearly defined identity, it becomes easier for him to avoid disordered behaviors. The male brain manages impulses better when guided by meaning and purpose.

My Personal Reflections

____

____

____

____

____

Endnotes:

1 Wu, J., Ou, X., Wang, X., Zilioli, S., Tobler, P. N., & Li, Y. (2022). *Exogenous testosterone increases sexual impulsivity in heterosexual men. Psychoneuroendocrinology, 145,* 105914. https://doi.org/10.1016/j.psyneuen.2022.105914

2 Guyton, A. C., & Hall, J. E. (2021). *Textbook of medical physiology* (14th ed., chap. 81, Male sexual function). Elsevier.

3 Barrett T, Tanner J, Gill AB, Slough RA, Wason J, Gallagher FA. The longitudinal effect of ejaculation on seminal vesicle fluid volume and whole-prostate ADC as measured on prostate MRI. Eur Radiol. 2017 Dec;27(12):5236-5243. doi: 10.1007/s00330-017-4905-x. Epub 2017 Jul 4. PMID: 28677052; PMCID: PMC5674119.

4 World Health Organization. (2021). *WHO laboratory manual for the examination and processing of human semen* (6th ed.). World Health Organization.

5 It is worth noting that there are some conditions that may alter this tendency to some extent, such as age, health conditions, emotional factors, and others.

6 Rodríguez-Nieto, G., Dewitte, M., Sack, A. T., & Schuhmann, T. (2021). *Individual differences in testosterone and self-control predict compulsive sexual behavior proneness in young males. Frontiers in Psychology, 12,* Article 723449. https://doi.org/10.3389/fpsyg.2021.723449

7 Centers for Disease Control and Prevention. (2025, September 24). *Sexually transmitted infections surveillance, 2024 (Provisional).* U.S. Department of Health and Human Services. https://www.cdc.gov/sti-statistics/annual/index.html

8 To expand on this point, consult Chapter 8 of the book *Dear Young Lovers...* by the same author.

9 Bates, D. (2025, March 27). *The impact of problematic pornography use in men. Psychology Today*. Retrieved from https://www.psychologytoday.com/us/blog/mental-health-nerd/202503/the-impact-of-problematic-pornography-use-in-men Psychology Today

10 Audiffren M, André N. The exercise-cognition relationship: A virtuous circle. J Sport Health Sci. 2019 Jul;8(4):339-347. doi: 10.1016/j.jshs.2019.03.001. Epub 2019 Mar 7. PMID: 31333886; PMCID: PMC6620209.

11 Audiffren M, André N. The exercise-cognition relationship: A virtuous circle. J Sport Health Sci. 2019 Jul;8(4):339-347. doi: 10.1016/j.jshs.2019.03.001. Epub 2019 Mar 7. PMID: 31333886; PMCID: PMC6620209.

12 Salmon, P. (2001) Effects of Physical Exercise on Anxiety, Depression, and Sensitivity to Stress: A Unifying Theory. Clinical Psychology Review, 21, 33-61. http://dx.doi.org/10.1016/S0272-7358(99)00032-X

13 Pargament, K. I. (1997). *The psychology of religion and coping: Theory, research, practice*. Guilford Press.

14 Koenig, H. G., King, D. E., & Carson, V. B. (2012). *Handbook of religion and health* (2nd ed.). Oxford University Press.

15 Newberg, A. B., & Waldman, M. R. (2009). *How God changes your brain.* Ballantine Books.

16 Smith TB, McCullough ME, Poll J. Religiousness and depression: evidence for a main effect and the moderating influence of stressful life events. Psychol Bull. 2003 Jul;129(4):614-36. doi: 10.1037/0033-2909.129.4.614. Erratum in: Psychol Bull. 2004 Jan;130(1):65. PMID: 12848223.

17 Guyton, A. C., & Hall, J. E. (2021). *Up cit.*

18 Brizendine, L. (2006). *The male brain.* Broadway Books.

19 Ohio State University. (2011, November 27). *Study debunks stereotype that men think about sex all day long*. Ohio State News. https://news.osu.edu/study-debunks-stereotype-that-men-think-about-sex-all-day-long/

20 This change appears in the DSM-5-TR, which is the diagnostic guide for psychologists and psychiatrists.

21 Hamann, S., Herman, R. A., Nolan, C. L., & Wallen, K. (2004). Men and women differ in amygdala response to visual sexual stimuli. *Nature Neuroscience, 7*(4), 411–416. https://doi.org/10.1038/nn1208

22 Gottman, J. M., & Silver, N. (2015). *The seven principles for making marriage work: A practical guide from the country's foremost relationship expert* (Revised and updated ed.). Harmony Books.

23 Ronghe V, Pannase K, Gomase KP, Mahakalkar MG. Understanding Hypoactive Sexual Desire Disorder (HSDD) in Women: Etiology, Diagnosis, and Treatment. Cureus. 2023 Nov 30;15(11):e49690. doi: 10.7759/cureus.49690. PMID: 38161863; PMCID: PMC10757759.

CHAPTER 6

MAN AND MONEY: LET'S TALK SERIOUSLY...

The content of this chapter is not intended to be considered financial advice. It is included in this material solely for educational purposes. If the reader finds it appropriate to consider any of the ideas presented, it is recommended that, before applying them, they consult their accountant or financial advisor.

The topic of money is of great importance to men. Due to a set of factors related to male psychology, cultural influence, and social formation, men often perceive money as an element closely linked to their identity and to the role they play.

The consumption demands of the modern family are numerous, and the necessary resources are not always available to meet them. In May 2025, the Federal Reserve System, in its report on the economic well-being of U.S. households, highlighted that 45% of parents with children under 18 would not be able to cover a financial emergency greater than $400 without resorting to debt.[1] If this is the reality in that country, we can imagine what is happening in other regions of the world with less economic power.

Male Psychology of Money

For men, talking about money is often a matter of security and control. Why? The reason is simple: money represents stability, the ability to plan ahead, and the capacity to respond to challenges—whether real or imagined. Male emotion is impacted by the presence or absence of money. Often, he thinks like this: "If I have money, I am fine, because I can solve personal or family problems." On the other hand, when he does not have it, his heart may fill with anxiety that he rarely recognizes or expresses, for fear of exposing his vulnerability. Instead of talking about the issue and seeking collective solutions, many men become irritable, distant, or immerse themselves in work overload, which over time can affect their family relationships and even their health.

Men often associate money with their identity and personal worth. Mistakenly, they conclude that if they can be great providers, then they are valuable men. If they cannot provide as they would like, they often see themselves as stuck, insufficient, or failures. It is important to clarify that we men are not numbers—even if we produce them. We are much more than that! Making judgments about your manhood, your competence, or your practical worth based solely on financial results can become a form of self-sabotage.

Money should be seen as a tool: a useful instrument that you can use, but one that must be treated with respect. It serves you, helps you solve situations, and facilitates the achievement of certain goals; however, giving it the responsibility of becoming the reason or foundation of your joys or sorrows can be extremely dangerous. If a man were to observe himself in relation to his money and his home, what thoughts would invade him? Some might say: "I have too heavy a financial burden," or "in this house everything—or everyone—depends on me." When money is not enough, the mature and wise man sees no problem in asking for help, reorganizing priorities, and readjusting strategies. After all, it is worth remembering that the family is, above all, a team.

From the male perspective, money is also often perceived as an expression of love. Perhaps this truth is never spoken aloud to the wife and children, but in that man's mind there is an underlying feeling like this: "If you see me working so hard, it is because I love you." "If I strive to keep all the bills paid, it is because you are important to me." "I want to guarantee your material well-being and that of the children." I can almost hear him say: "Because I am your husband (or your father): you will lack nothing." Does this style sound familiar to you?

From Man to Man...

Allow me to ask you three honest questions:

1. Do you truly believe that providing money is enough to say "I love you"? ____
   ______________________________.

2. If you continue down this path, what price will you end up paying... and what price will your family pay? ________
   ____.

3. If your income increases, but your emotional presence at home decreases, then... what are you teaching your family love really is? ______________________________

______________________________.

A Balanced View of Money

Contrary to what some believe, the Bible is not against money. Recently, I participated in a financial training where I heard the presenter say that, "according to the Bible, 'money is the root of all evil.'" Clearly, that is not what the Bible says. Rather, the text of 1 Timothy 6:10 was being used incompletely, which actually states: "For the love of money is a root of all kinds of evil..."

In this regard, it is important to highlight the following:

1. **God is not poor.** If there is anyone who understands money and possesses more resources than the entire list of billionaires in Forbes magazine combined, it is the God who created the heavens and the earth. Gold, silver, precious stones, and all the properties of the universe belong to Him. Psalm 24:1 says: "The earth is the Lord's, and all its fullness, the world and those who dwell therein."
2. **The ability to accumulate, manage,multiply,[2] and retain money [3] is a gift from God.** Properly combining these four areas of finance is truly challenging. The history of many who have studied finance makes it clear that time and good intentions alone are not enough to achieve it. The Bible declares: "And you shall remember the Lord your God, for it is He who gives you power to get wealth" (Deuteronomy 8:18).
3. **Clearly, the problem is not money, but the love of money.** Greed leads many into unlawful practices; however, we cannot blame money for the serious

condition of the human heart. Money is cold, insensitive, and obedient—it has no emotions.

4. **Human greed**—and the inability to determine how much is enough—can drag anyone into a relentless race without clear direction, only to end up empty, sick, and alone.
5. **The issue of money is, to a great extent, a matter of character.**[4] What is interesting is not simply making money—anyone can do that. What truly matters is allowing character to develop while we work to grow our finances.

Essential Principles for Good Financial Management

As men—whether single or married—the following principles and suggestions can help us bring structure and purpose to the management of our finances. We will divide this resource into three levels: basic, intermediate, and advanced. We will focus on general principles of personal and family financial management, character development, and a healthy worldview of money.

Basic Level

1. Goal: Develop awareness of the value of money, establish order, and cultivate self-control.

Clarifying:

- A man's money does not define his value—nor the amount he earns, nor how he manages it.
- Earning a lot of money does not necessarily mean that a man manages his resources well.
- What matters most is not how much a man earns, but how he manages what he receives—whether little or much.

- Money often exposes and amplifies a man's character; if there is order, it strengthens and multiplies it.

2. Learn to Work with a Budget

Why? Simple: money does not know how to find direction on its own. Therefore, the man who does not tell his money where to go will end up spending his time trying to figure out where it went.

Creating a budget is important, but following it requires discipline and consistency.

How is it done?

- Take a blank sheet of paper and write your real income according to its sources: salary, commissions, and others.
- Classify your expenses into three categories: fixed, variable, and discretionary.[5]

Important note: If you are going to spend even one dollar, make sure that dollar has a name and a last name (that is, a clear purpose).

A budget for someone who is a partner with God could be structured as follows:

- 10–15% Giving (tithes and offerings)
- 50–60% Basic household needs
- 10–15% Savings
- 20–30% Lifestyle

3. Create an Emergency Fund

Most people are not prepared for a major financial emergency, but this can happen to anyone, at any time. Three months is the suggested minimum; however, the ideal is to accumulate six months' worth of your monthly expenses.[6] If you are self-employed, it would be wise to consider at least twelve months.

What happens when a man does not have an emergency fund? He becomes more vulnerable to frequent episodes of financial stress, which can affect his mood, his family relationships, and his emotional leadership within the home.

4. Address the Issue of Debt

For better debt management, consider these general principles:

- If a debt cannot pay for itself, it is better to avoid it. Otherwise, it may enslave the one who is free.
- Understand that not all debt is the same: there is productive (good) debt and consumer (bad) debt.[7]
- Some experts recommend starting by paying off the smallest debts and then using that freed-up amount to strengthen the payment of the next. Others suggest eliminating debts with the highest interest rate first.

- Make sure to pay each debt on time, even if it is only the minimum payment.
- If possible, sell items you do not need to support debt repayment. Likewise, if you receive extra money, use it strategically to cancel or reduce a debt.

General Principles for the Intermediate Level

This level helps a man define a clearer vision for his personal and family finances. It allows him to work within a favorable structure to achieve better financial performance, with concrete goals, strategic decisions, and sustainable habits.

1. Goals

At this stage, the first thing you need to do is establish the goals you want to achieve. A financial life without goals is like a journey without a destination: energy, time, and resources are spent, but there is no forward direction.

It is important to define clear, precise, and measurable objectives in the short, medium, and long term. For example:

a) "Within the next six months, I will pay off $8,000 in credit card debt."

b) "I will prepare to purchase a home within one year, but without becoming financially suffocated. To achieve this, these are the steps I will take..."

c) "In five years, my children will enter college. This is what I plan to do to finance their education without loans."

2. Saving

Why save?

No matter how much you earn, today's consumer-driven society is designed for you to spend everything... or even more than you receive. That is why, in many cases, what you save is the only portion that truly becomes yours.[8]

When saving, consider the following:

a) **Do not save out of fear; save with purpose.**
Even if you could only save $50 per week, it would have a real impact: in one year it would be $2,600, and in ten years $26,000 (not including other possible gains or returns). What seems small today becomes significant tomorrow.

b) **Set aside a portion for investment projects.**
However, before investing your first dollar, inform yourself well. Evaluate carefully and ensure you have reasonable safeguards to protect your investment.[9]

c) **Protect yourself.**
If you want to grow financially, you need a solid foundation to support your financial structure. This includes insurance, and I want to highlight two in particular:

i) **Health insurance**
What took you decades to build can collapse in weeks—or even days—due to a medical crisis. In this sense, health insurance is like an umbrella that protects you in the middle of a storm.

ii) **Life insurance**
I understand that some men do not like to talk about this. However, I have never seen a man die prematurely simply because he had life insurance. What I have seen are households seriously affected by failing to plan for the one event we all inevitably face. For many, having this provision is an act of responsibility and love.

3. If You Are Married: Finances in Marriage

Now comes a key question: is it better to work together or separately?

The Bible says: "Two are better than one" (Ecclesiastes 4:9–12), and rightly so: teamwork allows you to advance further and better.

However, there are realities that cannot be ignored. When one spouse struggles with addiction (alcohol or drugs), wastes what they receive, or when money seems to "disappear," trust is seriously broken. In such cases, before attempting to manage finances as a team, it is wise to seek professional help to restore responsibility and stability.

Practical advice for married men:

a) Be transparent and trustworthy. This strengthens the bond and builds trust.

b) When making financial decisions, consult your partner and value their input.

c) Set aside 20–30 minutes each week to discuss household finances: review what was done, what is coming, long-term plans, contingencies, and goals.

4. Invest in Your Financial Education

I have heard many men say: "I would like to have a lot of money."

But then you discover that they have no plan to become financially educated. This truth must be stated clearly: Having money without financial education can be dangerous.

I have met people who had money but did not know they were poor. All they had was money. And because they did not know how to manage it, they soon found themselves asking: "What happened to my money?" In a short time, they ended up with more debt and more financial problems than before.

Therefore:

a) Learn how money works. Ask yourself: do you want to work for money your whole life, or would you like to learn how to make money work for you?

b) Other important topics every man should consider when making short-, medium-, and long-term plans:

- The role of inflation in the real value of money
- Before any investment: seriously calculate risk vs. return
- Financial strength depends not only on what you earn, but on your ability to multiply and preserve what you manage

Third Level – Advanced Principles

When a man begins to value more advanced principles for managing money, it is important to remember a basic truth: there is no virtue in trying to make money quickly. What comes easily and quickly often goes easily and quickly.

What does this mean? That developing a reliable system to multiply your financial resources is a process—not a stroke of luck.

Likewise, building a legacy requires defined purpose and sustained discipline. But here is the good news: that discipline, over time, is usually richly rewarded.

At this level, you may consider:

1) Study Investments: Diversity, Function, and Risk

Learning about investments involves understanding two inseparable elements: how they work and what risks they carry. Every financial move has a different level of risk, so it is important to consider your age, emotional profile, and personal tolerance for volatility.

The wisest approach is to educate yourself well and, when necessary, consult a professional who can guide you according to your goals and financial reality.

i) Real estate

Some men prefer to invest in real estate. The emotional stability it provides, along with certain tax benefits, are among the reasons many choose this path. With proper planning, even acquiring your primary residence may be more accessible than you imagine.[10] As a first-time buyer, you may also take advantage of programs offered by banks, municipalities, and support initiatives.

After 4–5 years, you could refinance or request a line of credit based on the property's equity and use those resources to acquire a second property as an investment.

The key point is this: what seems impossible today may become strategy tomorrow.

ii) Stock market

Others choose to invest in the stock market. This includes options such as stocks, index funds, ETFs, bonds, and mutual funds. At a more advanced level, there are options (calls and puts), trading (day trading, swing trading), and other derivatives.

However, investing in the stock market without proper education and sufficient practice—even simulated practice—can be like jumping into the void: it leads to losses, stress, and frustration.

A fundamental principle is this: increase your knowledge, set aside ambition, and always use reason. Move slowly. Do not invest money you are not willing to lose. And remember: the older you are, the less aggressive your investment strategy should be.

2) Learn to Manage Compound Interest

If we want to talk about real financial growth, we must talk about compound interest. What seems small in the present can become powerful over time.

Using the earlier savings example: if you open a compound interest account and deposit $50 weekly ($200 monthly) at an estimated annual rate of 8%, in ten years you could accumulate approximately $34,767.75.

You can explore online calculators and tables to visualize different scenarios. But remember: this type of project is built long-term. The more time you allow it to grow, the better it matures.

3) Indexed Universal Life Insurance (IUL)

Indexed Universal Life (IUL) insurance is a permanent life insurance policy that includes a cash value account linked to a stock market index. Although it is connected to market performance, many designs include protections that limit direct market losses.

Over time, this cash value can grow and be used as a loan, retirement supplement, estate planning tool, or family protection.

However, like any financial product, it has advantages and disadvantages:

- Higher costs associated with this type of policy
- Risk that if payments are low or inconsistent, the policy may weaken or collapse
- If loans are not managed properly, they may consume the accumulated cash value

In other words, this product does not work the same for everyone. It is usually more suitable for those with stable income and the ability to maintain consistent payments. Therefore, it should always be evaluated with a financial professional.

Let Us Consult the Bible and Complete the Idea

1. "For the love of money is ___________ ___________ ___________ ___________" (1 Timothy 6:10).
2. "A man's life does not consist in the ___________ ___________ ___________ ___________" (Luke 12:15).
3. "But if anyone does not _____ for his ___________, and especially for those of his own ___________, he

has denied the __________ and is worse than an __________" (1 Timothy 5:8).

4. "Who then is the ____________ and ___________ steward?" (Luke 12:42).

5. "It is more blessed to _______ than to ________" (Acts 20:35).

From Man to Man

Review the previous exercise. If you had to apply three major principles based on these biblical verses, what would they be?

____

____

____

How to Train the Male Brain for Good Money Management

When a man decides to train his brain to achieve financial stability and success, he seeks to strengthen self-control in managing his economic resources.[11] To achieve this, mental clarity and intelligent spending control[12] become essential.[13]

This man sharpens his vision—not only in the short and medium term, but also in the long term[14]—and is willing to discipline himself to reach his goals. His behavior benefits not only himself; it also becomes a living model of financial leadership within the family.[15]

Below are ideas you can consider for a brain-training plan aimed at promoting financially healthy masculinity. The program is designed for six weeks; it is practical, easy to apply, and adaptable whether you are single or married.

Exercise 1: Awakening My Awareness

My Goal

This week I will set aside the "it can't be done" mentality. I will focus not on blaming others for my situation, but on recognizing my reality and working from it. I am convinced that, with God's help, I will move forward!

What Will I Do?

I will work on three important exercises:

Financial audit (1 hour)
I will dedicate one hour to a complete audit of my finances, including: personal income, a detailed list of all debts (large and small), all fixed expenses, recent impulsive expenses, and an inventory of active subscriptions and memberships.

Expense tracking (without exception)
I will faithfully track every expense I make, no matter how small. At the end of the week, I will evaluate my list critically to determine what should remain and what can be adjusted—even temporarily.

Pause, ask, and buy
This week I will not make quick purchases. When I feel the urge to buy, I will pause for 90 seconds and ask:

- Do I know exactly how much I owe?
- Do I clearly understand the amount I am about to spend?
- Do I need it, or do I just want it?
- In one or two months, will it still be necessary?
- If I do not buy it, will anything truly serious happen?

Exercise 2: Maintaining Control

My goal

This week I will systematize my money. Since money has no

emotions, I will make sure that my emotions are not part of the equation either.

What will I do?
I will work on three exercises that will help me regain order and reduce financial tension:

I will create a budget.
I will include: income, fixed expenses, savings and investment, debt payments, and a specific amount for those "little cravings" that sometimes appear. Since God is my owner and partner, when preparing my budget, I will place Him in the position that corresponds to Him.

I will automate key deposits.
I will schedule automatic deposits for two areas:

- first, the minimum payment of credit cards or debts (if I have any), second, a portion designated for savings.
- I can start with 5%, but I will work to gradually increase it until reaching 15%.

Financial fast.
This week I will choose one day —and if possible two— in which I will not make any purchases, neither physical nor virtual.

Exercise 3: Impulses in Check!

My goal
I have noticed that, at times, stress leads me to spend. Without realizing it, this has become a pattern. This week I will strengthen my self-control and keep my impulses under control.

What will I do?
My task will be simple, but effective: I will replace quick reward with strategic reward.

Strategic transfer.
When I feel the desire to buy, I will try to avoid it. The money I was going to spend, I will transfer to my savings or toward paying a debt.

I will eliminate impulsive triggers.
I will identify two triggers that push me to spend. For example: online stores, shopping search engines, or ads on social media. This week there will be no online stores, no search engines, nor advertising pursuit for me.

My secret list of "temptations."
I will write 10 things that make me buy without thinking it through. Then I will design a practical plan to reinforce analysis before purchasing.

Exercise 4: The Man and His Debts

My goal

I commit to cancel all my debts and move from slavery to financial freedom.

What will I do?

This week I will strengthen delayed gratification and maintain the hope that, with God's help, things will improve step by step.

I will choose a strategy to pay debts.
I have several options, but I will choose one:

- **Snowball:** pay the smallest debt first while making minimum payments on the others; once one is finished, I transfer that amount to the next.
- **Avalanche:** attack first the debt with the highest interest rate.

I will define the payment frequency.
I will determine how much I will pay and how often: weekly, biweekly, or monthly, according to my income flow.

Cutting leaks.
I will identify money leaks: do I really need all my subscriptions? Can I eliminate one or two impulsive expenses and redirect those resources toward paying debts?

Exercise 5: Savings and Investment Mode

My goal
I will continue taking concrete steps toward a solid financial future. I want to strengthen my reserves and apply responsible multiplication strategies.

What will I do?
I will assess where I am today and project my vision forward.

I will build an emergency fund.
I can start with $500 or $1,000, but I will grow it until it represents three months of expenses... and then six.

I will begin to invest safely.
I will ask myself key questions:

- Am I contributing with my employer to a pension fund or 401(k)?
- Are my Social Security records correct?
- Is there any mutual fund, index fund, compound interest option, or another reasonable option that I can consider, once I have reliable information?

I will live below my status.
I will seek to spend less than I could spend. If I manage to reduce my expenses by 10%, my financial stability will increase and my capacity for saving and investing will be greater.

Exercise 6: A Faithful Steward

My goal
I want to forge a financially stable masculinity that honors God.

What will I do?
I will strengthen my identity in Christ and not in consumption. I will repeat the strategies practiced until they become solid, pleasant, and sustainable habits.

Weekly evaluation (20 minutes).
Each week I will set aside 20 minutes to review:

- What came in?
- How much went out?
- What unnecessary expenses did I have?
- What adjustments should I make this week?
- What will be my next goal?

I will write my financial profile.
I will ask myself deep questions:

- Who am I with money?
- What habits define my financial style?
- What things will I not do again?
- What system will I maintain?

Accountability.
I will seek someone trustworthy who can be my mentor in this area. I will talk with that person about my progress, ideally every two weeks.

A Lesson to Remember

Ali Hafed was a farmer, a landowner, satisfied and at peace with his life. One day, a visitor spoke to him enthusiastically about diamonds—explaining that people had become wealthy by finding them and that treasures were hidden beneath the earth in different regions.

That conversation stirred something deep within Ali. For the first time, he felt that what he had was not enough. Driven by ambition, he sold his farm and left his home in search of diamonds. He traveled, searched, exhausted himself, and became impoverished. Years passed, and his search turned painful. Eventually, overwhelmed by frustration, he died in misery.

Meanwhile, the man who bought his farm discovered something remarkable: shining stones in a stream crossing the land. They were diamonds. The treasure Ali had sought elsewhere had always been in his own land.

When your heart becomes consumed with the desire for wealth, never allow your relentless search for "something more" to cause you to abandon what is most valuable—the true treasure already in your hands: your family and your values.

My Personal Reflections

____

____

____

____

____

Endnotes:

1. Board of Governors of the Federal Reserve System. (2025, May 28). *Adults who would cover a $400 emergency expense using cash or its equivalent* [Data table]. Federal Reserve. https://www.federalreserve.gov/consumerscommunities/sheddataviz/unexpectedexpenses-table.html

2. Kiyosaki, R. T., & Lechter, S. L. (2017). *Padre rico, padre pobre: Qué les enseñan los ricos a sus hijos acerca del dinero* (Ed. actualizada). Aguilar.

3. Kiyosaki, R. T. (2011). *El cuadrante del flujo del dinero: Guía del padre rico para la libertad financiera.* Aguilar.

4. I heard this concept from Professor Pedro H. Morales at a conference on finance for business men and women, held in New York City in the summer of 2018.

5. Discretionary expenses are those you incur when using the money you have left after meeting your obligations. It is important to keep track of them for better financial management.

6. Ramsey, D. (2008). *La transformación total de su dinero: Un plan efectivo para alcanzar bienestar económico.* Grupo Nelson.

7. Hiciano, W. (2014). *Hablemos de dinero: Cómo mejorar sus finanzas personales.* IADPA.

8. Housel, M. (2024). *La psicología del dinero: 18 claves imperecederas sobre riqueza y felicidad.* Booket.

9. Burkett, L. (1996). *Usando su dinero sabiamente: Using your money wisely* (1.ª ed.). Editorial Unilit.

10 Morales, P. H. (2017). *Verdaderamente rico* (6.ª ed.). Editora FR Multiservicios.

11. Bai R. Impact of financial literacy, mental budgeting and self control on financial wellbeing: Mediating impact of investment decision making. PLoS One. 2023 Nov 14;18(11):e0294466. doi: 10.1371/journal.pone.0294466. PMID: 37963159; PMCID: PMC10645357.

12. Devis, C. (2021). *Un inmueble al año no hace daño.* Independently published.

13. de Almeida, F., Scott, I. J., Soro, J. C., Fernandes, D., Amaral, A. R., Catarino, M. L., Arêde, A., & Ferreira, M. B. (2024). Financial scarcity and cognitive performance: A meta-analysis. *Journal of Economic Psychology,* 101, 102702. https://doi.org/10.1016/j.joep.2024.102702

14. Adee, S., Lau, E., & Wong, S. (2024). Mediating factors in retirement savings and well-being: a focused study on Sarawak. *Cogent Economics & Finance,* 12(1). https://doi.org/10.1080/23322039.2024.2422217

15. Spoelma TM, Fletcher KA. Financial stress and leadership behavior: The role of leader gender. J Occup Health Psychol. 2024 Oct;29(5):317-341. doi: 10.1037/ocp0000387. PMID: 39388139; PMCID: PMC11892524.

CHAPTER 7

THE APPARENT FEMALE MYSTERY

He was a hardworking man, serious and few of words. He did not know how to name what he felt, much less read what others felt. He grew up learning to "solve" life with strength and money: paying bills, maintaining a roof, providing food, offering security. And when he met that woman—beautiful on the outside, but even more beautiful within—he thought God had given him a miracle. She possessed nobility, patience, a willingness to build a home, and a loyal love that gave itself with hope. Even with certain insecurities, she chose to trust him. They married, built a decent home, and from the outside appeared to be a stable story. He loved her... in his own way. And for him, loving meant fulfilling, bringing, sustaining, protecting.

But she did not need only provision; she needed presence. She needed eye contact, conversation, interest, tenderness, validation. She needed to feel seen. And that, he did not understand. It was not bad intention; it was emotional illiteracy. He did not detect signals. When she said, "I feel alone," he responded, "but you lack nothing." When she remained silent out of exhaustion, he celebrated it as peace. When she tried to talk about her day, he answered with monosyllables or changed the subject. They lived in the same house, but little by little stopped inhabiting the same world. She began to feel like a shadow that cooks, cleans, smiles—but does not exist. And over time, she stopped asking, because every request ended in frustration. The deterioration was slow and silent, like moisture inside a wall: first the soul stains, then trust cracks, and finally affection collapses. What was warmth became routine; what was desire became distance; what was home became a cold place.

He noticed the change too late. He noticed that she no longer laughed as before, that she no longer prepared herself with enthusiasm, that her hugs were brief, that her gaze lived tired. But he interpreted it as "phases," or "hormones," or "women's things." Meanwhile, she had spent years in an emotional waiting room: waiting for a gesture, a question, an apology, a caress that did not ask permission, an "I understand you" that did not come with a sermon. She tried in many ways: she spoke, cried, argued, wrote messages, stayed silent, became more accommodating, then stronger... until her emotional tank was empty and hollow. And when a woman reaches that point, she no longer protests: she shuts down. She no longer fights for love; she grieves it in silence until she buries it.

One day, without shouting or drama, she told him: "I am no longer interested in continuing this relationship." He froze. He offered what he always had: "But I have given you everything... you lack nothing." She took a deep breath and replied with a calm that hurt more than a shout: "What I lacked was the most basic thing: to feel loved. Not loved through your payments, but through your heart. I got tired of competing with your phone, your silence, your exhaustion, your indifference. I lived for years next to a man who did not know how to recognize my presence at home. I was there, but I was invisible. I waited for emotional crumbs, I waited for you to look at me, to listen to me, to choose me in the small things. And it never happened. My tank is empty. I feel nothing anymore." He begged, promised change, said he would learn, that he would go to therapy, that he would do whatever it took. She simply lowered her gaze: "It's not that I don't believe you... it's that it's already too late for me. I don't want to come back. There's no way."

And then he understood something terrible: love also dies of starvation. That night, in his room, for the first time he felt the weight of his emotional poverty. He was not losing a wife; he was harvesting years of absence. And he broke down—not from guilt, but from awakening. Over time, he sought help, allowed himself to be taught by a counselor, learned to speak, to identify emotions, to ask for forgiveness without excuses, to listen without defending himself. But she did not return. Their story did not have a romantic ending; it had a sobering one. She saved herself. And he, with tears, saved himself late... but he saved himself. Because he understood, for the first time, that providing does not replace connection; that a home without affection is only a building; and that love is not sustained by silent intention, but by active presence.

Years later, he was no longer the same. Where once there was a "strong" and closed man, now there was a man with a trained heart. He learned to be emotionally literate: to name what he feels, to recognize what the other needs, to care before losing. And although he carried the pain of an opportunity destroyed by his ignorance, he transformed his failure into a message for other men: "Do not wait for love to die to start loving well." Because there are losses that cannot be repaired... but they can become lessons that save other homes.

Are Women Difficult?

For some men, women are curious beings who combine mystery with complexity—someone difficult to understand. A true project... Some have even suggested that the man who listens to her, understands her gestures, interprets her correctly, and responds accordingly deserves a trophy. Are they right to think this way? For now, allow me to set aside that discussion and focus on another fundamental aspect.

Long before taking his first breath, every man already lived his first experience with a woman. From the moment he was conceived, his mother carried him in her womb. Throughout the entire gestational process, there was a relationship between them that was not accidental or random, but a close and profound connection. The mother's physical and emotional well-being directly influenced the baby's development: both in the small brain being formed and in how that child would later learn to respond to stress.

From there, in a man's life, a woman plays different roles, each with its own characteristics. She is not only a mother; she is also a sister, a friend, a wife, and a daughter.

As a mother, she becomes her son's first emotional world: the first female figure who embraces him, welcomes him, and gives him his first lessons in tenderness, love, emotional language, boundaries, security, nourishment, and comfort. The degree of emotional connection a child experiences with caregivers—particularly with his mother—can influence the quality of early attachment; and this, in turn, often affects how he will establish and maintain social and romantic relationships in adulthood.[1]

For a man's emotional development, the sibling environment holds a value that is often underestimated. A sister is not viewed from a romantic bond nor from a maternal role: she is a female

presence in the realm of the fraternal, where closeness is built without seduction and coexistence occurs without possession. This setting is ideal for cultivating respect, collaboration, empathy, and healthy protection free from control. In this dynamic, a man learns to see a woman as his equal: with dignity, as a complete person, without the need to possess or use her.

And then an inevitable question arises: what happens when a man did not have sisters? An only son—or a boy raised only among males—may face greater difficulty relating to women, not out of bad intention, but due to lack of emotional exposure. Unlike someone who had a female sibling as a "training ground," some men reach adulthood without having practiced, in daily life, the art of coexisting with feminine sensitivity from a safe and respectful bond.

From the role of a friend, when a man relates to a woman, he learns to see her beyond someone he can use or with whom he shares a romantic relationship. Healthy friendship creates a space that serves as a school for cultivating respect, empathy, maturity, good communication, and the establishment of boundaries. And why not? It can also help prepare him for a future relationship.

However, before a man, a woman is especially distinguished by her role as a wife. Among human relationships, this is one that offers the greatest potential for emotional growth.[2] And here it is worth stating a truth without embellishment: growth rarely occurs without pain. Nevertheless, pain is not sufficient reason to abandon development. The bond with a wife is particularly demanding for a man, because she is not simply his housemate or domestic partner. She builds him, yes; but she also reveals him. She exposes him. She brings to light what is inside him, confronts him, and in that process, if he allows it, forces him to mature and grow emotionally.

Then there is her role as a daughter, which produces a unique impact on a man. The way a father treats his daughter nourishes her emotional security, while her presence often awakens in him a stronger protective instinct than that which emerges with a son.[3] A daughter teaches him a tenderness with purpose: patience, compassion, and sensitivity that does not weaken him, but humanizes him. When she honors, listens to, and respects her father, she strengthens his sense of role and gives him deep satisfaction. But contempt or disrespect can fracture the bond and deeply wound his masculine identity.

At this point, the initial question still remains: are women difficult? In light of what has been presented, the answer cannot fit in a single syllable. More than difficult or mysterious, they are different. That difference is not explained only by biology, but also by the formative context that shapes them and by that set of invisible elements that make up the male emotional "radar." In other words: understanding a woman is not a matter of guessing, but of vision; not of trophies, but of maturity.

Myths and Realities of Men and Women

For several years, I have had the privilege of working with men from multiple cultures and different lines of thought. Along that journey, I have heard phrases repeated over and over as if they were absolute truths. Below, I share—in a format of myths and realities—some common ideas that often circulate in the male mindset.

1) "This woman is very intelligent; you have to be careful..."

Myth: *Female intelligence is a threat to my authority as a man.*

Reality: Female intelligence is a valuable resource that can contribute to building a better home. This is not about competing, but about sharing. A brilliant woman does not come to dim your light; she comes to illuminate areas you were not seeing. She can reveal blind spots, clarify vision, and help reach the goal more safely. If you have a brilliant woman, remember this: her brightness does not diminish yours; it can elevate you.

2) "If my wife has a car, a job, and a visa... I've lost her."

Myth: *If she is independent, it means she no longer needs me.*

Reality: In some Latin American countries, I have heard the idea: "If a woman has a car, a job, and a U.S. visa... you've lost her." Is this true? The fact that she can drive, works, and travels does not mean she wants to be alone. It means she can choose. And if she chose you, it was not out of necessity—it was out of love. A wise man is not unsettled when his wife is capable; on the contrary, he finds rest. Because a woman who is able, and still chooses to stay, stays out of conviction.

3) "My wife cannot earn more than I do. That's a serious problem."

Myth: *If I earn more, I am worth more.*

Reality: A man's worth and his leadership in the family are not measured by the amount of his income, but by the quality of his character. It is measured by his vision, by his sense of responsibility, by the discipline that defines him, and by the emotional protection he provides. And in a very special way, it is measured by his ability to model Jesus at home. The one who earns more is not the greater leader; the greater leader is the one who knows how to love better and can show it without it being a trade-off.

4) "If she is more academically prepared... it's risky."

Myth: *Her education will lead her to arrogance or infidelity.*

Reality: Her education is a shared asset. The real issue is not her growth, but the man's insecurity.

5) "She talks too much about emotions... she's attacking me."

Myth: *Talking about feelings creates conflict.*

Reality: For a woman, expressing emotions is usually a natural way to connect, not a strategy to destroy. When she speaks to you from her emotional world, she is often not asking you to fix things; she is asking you to accompany her. Many times, all she needs is to be listened to with attention and empathy. If you learn to listen without becoming defensive, you will see something remarkable: the conversation stops being a war and becomes a bridge.

6) "If she points out my mistakes, she doesn't respect me."

Myth: *A woman cannot correct a man.*

Reality: When a woman still corrects you, it is because you still matter to her. Deep down, she is saying: "I want you to be better, because your life matters to me." Female indifference is worse than complaint. If your woman has stopped talking, should you be concerned? At times, it could be a sign of emotional disconnection. Therefore, blessed is the man who has by his side a woman who—instead of giving up—still fights for the growth of both.

7) "If she asks for boundaries and transparency, she wants control."

Myth: *Boundaries equal control.*

Reality: In a healthy relationship, agreements do not enslave; they stabilize. Transparency—whether in finances, in the emotional realm, or in the digital world—should not be seen as a prison, but as a style of organized trust. Boundaries are not chains; they are fences that protect what is valuable. And what is valuable in a marriage is peace, respect, and faithfulness.

8) "If she doesn't want intimacy, it's over—she doesn't love me."

Myth: *Love equals immediate sexual availability.*

Reality: In most cases, female sexual desire is not triggered by the visual, but by connection. It is nourished by the emotional security she experiences with her partner. Many times, the lack of intimacy is not due to a lack of love, but to a lack of attention, lack of details, accumulated fatigue, or poorly managed stress. In other words: often, her body is not saying "I don't love you," but "I am exhausted, disconnected, or emotionally alone."

A capable woman is not a threat; she is a valuable resource that, if I use it wisely, has the magnificent potential to make my life easier and more manageable. A brilliant woman is not an enemy, nor a mere decorative presence. She is an ally whose light can broaden my vision, help me better understand the landscape, and prepare us more clearly for the future we will have to walk together.

A woman with resources does not come to replace me; she comes to complement me. Even more: her presence can push me to discover the hidden nature of the man who lives within me, the one who sometimes hides behind ego, insecurity, or fear. What she has does not define my manhood; on the contrary, it reveals it, elevates it, and multiplies it.

From Man to Man: Reflection Exercise

After considering the myths mentioned above, take a few minutes to reflect honestly on your own way of thinking and acting.

1. Personal Identification

Of the myths presented, which one most closely reflects a belief, attitude, or practice you have adopted—consciously or unconsciously—in your life?

What experiences, models, or influences may have contributed to your thinking in that way?

---

2. Critical Analysis

When contrasting that myth with the reality presented in the study, what aspects of your way of thinking or acting do you recognize need to be reviewed or adjusted?

What consequences could result from maintaining that belief without questioning it?

---

3. Decision and Intentional Change

If you decided to take a concrete step to transform that belief or practice in your daily life, what specific action would you begin to implement today?

How could that change positively impact your character, your relationships, and your purpose as a man?

Male Emotional Connection and Disconnection

Let us talk a bit about male emotional connection. In practical terms, we can define it as an internal bond that allows a couple to experience closeness, security, and attunement. It is easy to recognize when both interact naturally: when one feels seen by the other, when dialogue flows, when there is space to feel valued, cared for, and supported emotionally.

It is not about momentary romanticism, nor merely coexisting in the same physical space. Emotional connection is, rather, the intimate conviction that the other person is there for you. It is being able to recognize—even in silence—this essential truth: "You are with me, and my heart is with you."

In real terms, male emotional connection:

- **Is expressed in the security he conveys to his partner,** communicating—through actions and words: "I am with you, not against you."
- **Is manifested in shared attunement and empathy**, as when she can perceive: "You understand me, and I truly matter to you."
- **Is evidenced in consistent response**, because there is a reliable dynamic: "When I reach out, you respond."
- **Translates into identity without fusion**: feeling that I remain "myself," fully aware of who I am, closely connected to you, yet without losing myself in you.

Attachment Theory: A Psychological Perspective

From a psychological standpoint, emotional connection is remarkably explained through **secure attachment.**[4] Simply put, attachment is the way a human being learns to connect with others in order to feel safe. It is a deep and lasting emotional bond that develops primarily between the baby and their main caregivers—father and/or mother—and fulfills a transcendent function.

Within that early bond lies the foundation of emotional security: the ability to regulate stress and fear, as well as the necessary basis to explore the world with confidence.[5] In other words, secure attachment not only teaches how to love; it teaches how to feel protected, to interpret closeness as refuge, and to build relationships where the heart can rest.

The British psychiatrist and psychoanalyst John Bowlby is considered the father of modern attachment theory. Another key figure was Mary Ainsworth, whose most important contribution was the development of the Strange Situation Procedure (SSP), which allowed systematic observation of how a baby bonds with their primary caregiver.[6]

Together with her colleagues, Ainsworth conducted longitudinal studies[7]—first in Uganda and later in Baltimore, United States—leading to the identification of three main attachment[8] patterns:

Secure Attachment

In this pattern, the child develops trust in the presence of the caregiver. They may experience distress when the caregiver leaves, but calm down relatively quickly upon their return. Consistent and sensitive care trains the child to regulate emotions safely, strengthening their capacity to trust.

Insecure Avoidant Attachment

When caregivers are emotionally distant, rejecting, unavailable, or constantly busy, the child learns— as a survival strategy— not to need. They display apparent independence and avoid close contact. Although they may seem calm, their emotional system remains activated, and during reunion they minimize connection rather than seek comfort.

Insecure Ambivalent/Resistant Attachment

In this pattern, the child experiences high distress when the caregiver leaves. Upon return, behavior is contradictory: they desire closeness but simultaneously resist it, showing irritability. They may take longer to calm down, as if their emotional world struggles to stabilize even when the caregiver is present again.

How Do These Patterns Translate into Adulthood?

Children who developed a secure attachment tend to become confident and trustworthy adults. They generally show a good capacity for communication, emotional stability, and a healthy willingness to become intimate with their partner without fear of losing themselves in the relationship. They know how to draw close, receive affection, and build bonds where security is not negotiated.

In contrast, when the attachment pattern was insecure-avoidant, there is a high probability that the adult will appear emotionally disconnected. He usually prioritizes independence as a refuge, avoids exposing his vulnerability, and develops a kind of "emotional self-sufficiency" that, although it appears to be strength, is often a defense mechanism. For this man, admitting emotional need can feel like weakness, and opening his heart can be deeply uncomfortable.

Furthermore, when the early pattern was insecure ambivalent/resistant, the adult usually experiences intense anxiety in the face of the possibility of abandonment. Fear—and sometimes dread—is triggered at the slightest sign of emotional distance. There then arises the need to monitor "everything": signals, words, silences, mood changes. Interestingly, although he desires closeness, he often does not manage to enjoy it; he suspects it, resists it, or sabotages it, as if love were something he longs for but does not fully believe he deserves.[9]

Invisible Wounds and Real Consequences

After Genesis 3—when the entrance of sin into the world is described—its effects were so devastating that there is not a single home that has remained completely unscathed. Therefore, even with the best intentions of our parents or caregivers, it is common to find emotional deficiencies and even subtle neglect that, over the years, leave deep wounds.

Hence, we see kind, well-mannered, and attractive men, but with serious difficulties when they try to connect emotionally with a woman. Are they destined to live single all their lives? No. Many marry and form a family; however, the trace of the avoidant or ambivalent/resistant pattern follows them, limiting their ability to crystallize the dream of a stable and functional home.

He is a good man: hardworking, responsible, with a noble heart. The problem is not his lack of values, but his lack of emotional awareness. He has not known how to name his wounds or interpret them. On the outside he seems to be fine; everyone sees him as happy, strong, and stable. But on the inside he knows that something is missing: he perceives that his relationship lacks deep connection and recognizes—although he does not always say it—that this has been a repetitive pattern.

Let Us Consult the Bible and Complete the Idea

1. "Two are ________ than ________; because they have a better reward for their labor..." (Ecclesiastes 4:9–10)
2. "The ____________ woman builds her ___________..." (Proverbs 14:1)
3. "He who finds a ___________ finds a good thing..." (Proverbs 18:22)
4. "Do not rebuke a _____________, lest he hate you; rebuke a ________, and he will love you." (Proverbs 9:8)
5. "A cord of three strands is not quickly __________." (Ecclesiastes 4:12)
6. (Love) "bears all things, believes all things, hopes all things, endures all things." (1 Corinthians 13:7).

How to Train the Male Brain to Overcome Attachment Patterns

Training the male brain to overcome avoidant attachment and ambivalent (anxious/resistant) attachment is a possible mission. Below, we will present practical resources that can help in that process. However, it is important to note that some readers may need professional support, especially if there are deep wounds, relational trauma, or persistent symptoms of anxiety and depression.

A fundamental truth that we must highlight is this: attachment is a neuroemotional pattern that is learned early in life, and it can also be reprogrammed. The brain, the nervous system, and the relational habits that are repeated over the years all play a role in the formation of this pattern. Researchers and clinicians have observed that, when a person undergoes a training process—a kind of emotional "brain gym"—significant changes can be achieved.

For that reason, we will present a six-week action plan that, although it can provide real benefits, is not intended to be a magic formula nor to replace therapy. Its purpose is to function as an educational tool: an initial map that motivates the reader to begin their process of growth and, if appropriate, to seek individualized help.

What Happens in the Brain of a Man with Insecure Attachment?

When a man's attachment is not secure, his brain tends to fall into two automatic modes.

1) Emotional deactivation mode: avoidant attachment
In this mode, the man "shuts off" his emotions to protect himself. His defensive style is activated. For that reason, his behavior often reflects coldness, distancing, and avoidance of commitment. He processes almost everything through the exclusive filter of reason. He often resorts to humor

as a sophisticated strategy to avoid his own inner reality. He works intensely, stays busy, and in the sexual area may focus more on gratification than on emotional connection. Without saying it, his behavior declares to the world: "I am fine. I don't need anyone." But deep down, that apparent independence is usually a mechanism: it is not the absence of need, but fear of needing.

2) Emotional hyperactivation mode: ambivalent attachment (anxious/resistant)

In this other mode, the brain turns on into "emotional alarm." This man insists, demands, and protests as someone who lives internally activated. He demands affection, attention, and constant signals of security. He wants to have everything under control. His jealousy betrays him. And when he cannot demand, he begins to ruminate: his mind repeats scenes, interprets signals, imagines scenarios, and becomes emotionally exhausted. He frequently feels the need to check his partner's phone, suspects easily, and his relational dynamic becomes filled with repetitive dramas. Each of these behaviors, on its own, shouts the same internal message: "They are going to abandon me... please, 911!" It is not malice: it is fear. And fear seeks control.

A Practical Training Program

Now then: let us look at a practical, simple, and progressive program designed to begin the process of emotional reprogramming.

Exercise 1: Diagnosis — Identifying triggers

My weekly focus: I will try to detect my pattern in real time.

Exercises:

- At the end of the day, I will take 5 minutes to self-evaluate:
- What emotion did I avoid or exaggerate today?
- What made me feel afraid?

- What made me feel rejected, humiliated, or abandoned?
- What did I do in response? Did I run away, remain silent, try to control, or start to complain?

Principle of masculine reflection:
If you cannot identify it, you cannot change it.

Exercise 2: Training the nervous system — Who commands whom?

My weekly focus: A man cannot truly change if his organism is in alarm. If the body commands, the mind is a servant.

Exercises:

This week I will do a daily 10-minute routine, divided as follows:

- 5 minutes of 4–6[10] breathing (inhale 4, exhale 6).
- 5 minutes of body scanning[11] to discover where tension is located: stomach, neck, chest, jaw, hands, etc.

Detecting tension early is crucial. This way you can work to release it before it explodes... or before you choose to isolate yourself.

This exercise helps the anxious person lower impulsivity, while it helps the avoidant person increase emotional tolerance.

Principle of masculine reflection:
With God's help I can train my prefrontal cortex and restrain my negative emotions.

Exercise 3: Reprogramming the emotional GPS

My weekly focus: I will take positive actions to advance in my emotional growth.

Exercises (according to your attachment pattern):

- If your pattern is avoidant, the goal is to stop "running away" affectively.

- If your pattern is ambivalent (anxious/resistant), the objective is to stop acting as a pursuer or controller.

A) If your pattern is avoidant:
Once a day, at the appropriate moment, you will express a real emotion in 20 seconds. For example:

- "Today I didn't see you arrive early. I was worried about you, although I didn't show it."
- "Look, I know it's hard for me to talk about this... but it's important that I say it."
- "Honestly, I tell you: I feel afraid that this is not going to turn out well."

This exercise opens the door to your emotional world, trains your courage, and promotes vulnerability... but in a measured way.

B) If your pattern is ambivalent (anxious):

Adopt this statement as a general rule:

"In alarm mode there are no complaints."

This is what I will do when I feel urgency (before exploding):

I calm myself (I perform the breathing exercise suggested above).

- I write the message, but I DO NOT send it.
- I wait, in silence, 20 seconds.
- This increases my emotional maturity and strengthens my tolerance for uncertainty.

Principle of masculine reflection:

To establish meaningful connections I must move from "threat" to "bond."

Exercise 4: Functional masculine communication

My weekly focus: I will use language typical of secure attachment and grow in that direction.

Exercises:

When I face conflicts or needs, I will use a simple three-sentence protocol:

1. **What I feel**: "I feel..."
2. **What I need**: "I need..."
3. **What I propose**: "Would it be okay if we try...?"

Practical example:

Suppose you are going through a difficult situation with your wife. Both are attending to your own agendas and have lost shared space. How do you address this without ignoring it or exploding?

____

____

____

____

____

If you are single, imagine the following situation: you are fulfilling your responsibilities and outwardly everything seems to be going well, but you feel internally disconnected. Your emotions are not clear and you perceive a noticeable emptiness within you, which is reflected in your way of interacting with others. How do you then address this moment with clarity, self-control, and direction, without denying it or overflowing?

____

____

____

As a man you could say:

- "I feel disconnected."
- "I need us to set aside a moment to talk."
- "Would it be okay if we take 15 minutes tonight, without screens, just you and me, to talk a bit?"

Principle of masculine reflection:
Silence and withdrawal are not viable solutions. Neither are complaint and protest in the wrong way.

Exercise 5: Immunity to conflict

My weekly focus: When conflict arises, I will tolerate closeness without fleeing or attacking.

Exercises:
A large number of couples have topics they cannot touch without provoking a "mini third world war." This week we will have a difficult dialogue, but with a clear protocol.

Duration: 20 minutes

Rules:

- No one may threaten.
- Sarcasm is out of place.
- There can be no total withdrawal.
- If tension rises, we take a 10-minute break and then return to the topic.

This exercise strengthens the man's emotional self-control when addressing sensitive topics. It also develops effective perseverance in matters of common interest.

Principle of masculine reflection:

Secure attachment does not mean absence of conflict, but presence of skills and willingness to repair the breach.

Exercise #6: Behaviors that reflect secure attachment

My weekly focus: I will strengthen measurable habits that promote secure attachment, remembering that the brain does not learn only through ideas, but through repetition.

Exercises — Habits I will cultivate this week:

1. **Connection ritual**: I will take 10–20 minutes each day to ask questions and listen actively.
2. **Intentional affection**: I will give at least one warm hug (if applicable) that lasts around 20 seconds.
3. **Weekly date**: Without screens, 60–90 minutes.
4. **Quick repair**.[12] If there was a conflict and the connection was interrupted, I will ask for forgiveness before going to sleep.

Making these practices a habit trains the couple in emotional security. In addition, it gives clear signals of predictability and increases trust. One of the greatest challenges of the human being is to establish healthy connections, because connection is biologically powerful. It has been documented that social pain[13]— produced by rejection or disconnection—shares neural bases with physical pain. This helps to understand why emotional disconnection weighs so heavily on both women and men.

A person who experiences social pain is not a mysterious character: he is a wounded human being who needs to be healed.

Principle of masculine reflection:

When a man grows emotionally, everyone at home benefits.

How can a man know if he is healing his attachment?

- A man can notice that he is healing his attachment when: He can talk about emotions without feeling shame.
- In threatening situations he no longer freezes or explodes.
- He can ask for what he needs without manipulating.
- He tolerates distance without automatically concluding that he has been abandoned.
- His wife can perceive that he is truly present.
- The cycles of pursuer and withdrawer have become episodes of the past.

How did he achieve it?

Training techniques have their place and serve an important purpose, but there is something even greater: the power of Jesus, who still today continues healing wounds and transforming hearts. He said: "Come to Me, all you who labor and are heavy laden, and I will give you rest" (Matthew 11:28).

An uncountable number of men found inner fullness by knowing Him. They discovered that the woman is not a mystery, but a gift from God: a suitable helper (Genesis 2:18), so that together they may reach the supreme purpose of life.

My Personal Reflections

_____

_____

_____

_____

_____

Endnotes:

1. Dinero RE, Conger RD, Shaver PR, Widaman KF, Larsen-Rife D. Influence of family of origin and adult romantic partners on romantic attachment security. J Fam Psychol. 2008 Aug;22(4):622-32. doi: 10.1037/a0012506. PMID: 18729676; PMCID: PMC2689376.

2. Messina I, Calvo V, Grecucci A. Attachment orientations and emotion regulation: new insights from the study of interpersonal emotion regulation strategies. Res Psychother. 2023 Dec 31;26(3):703. doi: 10.4081/ripppo.2023.703. PMID: 38224213; PMCID: PMC10849076.

3. American Psychological Association. (2017, May 25). Fathers' brains respond differently to daughters than sons. https://www.apa.org/news/press/releases/2017/05/fathers-daughters

4. Brandão T, Matias M, Ferreira T, Vieira J, Schulz MS, Matos PM. Attachment, emotion regulation, and well-being in couples: Intrapersonal and interpersonal associations. J Pers. 2020 Aug;88(4):748-761. doi: 10.1111/jopy.12523. Epub 2019 Nov 15. PMID: 31674659; PMCID: PMC7383855.

5. Flaherty SC, Sadler LS. A review of attachment theory in the context of adolescent parenting. J Pediatr Health Care. 2011 Mar-Apr;25(2):114-21. doi: 10.1016/j.pedhc.2010.02.005. Epub 2010 May 1. PMID: 21320683; PMCID: PMC3051370.

6. Van Rosmalen L, Van der Veer R, Van der Horst F. AINSWORTH'S STRANGE SITUATION PROCEDURE: THE ORIGIN OF AN INSTRUMENT. J Hist Behav Sci. 2015 Summer;51(3):261-84. doi: 10.1002/jhbs.21729. Epub 2015 May 19. PMID: 25990818.

7. Duschinsky, R. (2020). Mary Ainsworth and the Strange Situation Procedure. En T. (Ed.) Cornerstones of attachment research. Oxford University Press. https://doi.org/10.1093/med-psych/9780198842064.003.0002

8. Ainsworth, M. D. S., Blehar, M. C., Waters, E., & Wall, S. (2015). Patterns of attachment: A psychological study of the Strange Situation (Classic ed.). Psychology Press. https://mindsplain.com/wp-content/uploads/2021/01/Ainsworth-Patterns-of-Attachment.pdf

9. Fraley, R. C. (n.d.). Self-report measures of adult attachment. University of Illinois Urbana-Champaign, Department of Psychology. https://labs.psychology.illinois.edu/~rcfraley/measures/measures.html

10. This exercise helps reduce anxiety and the level of alarm in the body. Studies have demonstrated the effectiveness of this technique in helping to balance the sympathetic and parasympathetic response, facilitating a reduction in stress and anxiety.

11. Luo, Q., Li, X., Zhao, J. et al. The effect of slow breathing in regulating anxiety. Sci Rep 15, 8417 (2025). https://doi.org/10.1038/s41598-025-92017-5

12. Brittle, Z. (2014, September 3). R is for repair. The Gottman Institute. https://www.gottman.com/blog/r-is-for-repair

13. Eisenberger NI. The neural bases of social pain: evidence for shared representations with physical pain. Psychosom Med. 2012 Feb-Mar;74(2):126-35. doi: 10.1097/PSY.0b013e3182464dd1. Epub 2012 Jan 27. PMID: 22286852; PMCID: PMC3273616.

CHAPTER 8

INTEGRALLY FIT

Is it possible that some man feel trapped between two great pressures: the ability to be a good provider or someone who can demonstrate to everyone that his success is more than evident? In this invisible struggle, some men lose the sense of the things that matter most in life. What I truly desire and what justifies my direction are matters that deserve quiet consideration, which is the purpose of this closing topic of this journey.

In the year 1943, the American psychologist Abraham Maslow presented a proposal on the theory of motivation[1] that, over time, would be widely developed. Professor Maslow came to be recognized for his work, in which he highlights how people are stimulated and toward where their sense and purpose are directed, according to their needs. This explains a tendency of human nature in which purpose is determined, or at least strongly influenced, by interests.

Maslow did not focus on exploring or developing the theories previously presented by Skinner in 1938[2] and later in 1953.[3] Nor was he based on the findings of Freud[4] and later his work published in 1961,[5] related to behaviorism and psychoanalysis, respectively. His proposal appears to be a reaction to these approaches, and he presents, through a pyramid, what he called the "hierarchy of human needs."

The model proposes five levels, organized as follows:

1. **Physiological needs**: Food, sleep, health, sex and rest.
2. **Safety needs**: Protection, job and financial stability, order, health.
3. **Belonging or affiliation needs**: Family, friendship, partner, acceptance.
4. **Esteem or valuation needs**: Respect, recognition, achievement, reputation.
5. **Self-actualization needs**: Potential, purpose, mission.

Although human needs help us understand many of our motivations, the true purpose of the human being is not limited simply to satisfying them. The Bible offers a deeper vision, teaching that the meaning of life is found in the relationship with God and in the calling to reflect His character.

The needs presented may not be pursued in a strictly linear manner; at times, a person may focus on satisfying more than one at the same time. Likewise, it is possible that they are not always followed in the suggested pyramidal order. Regardless of the order or the time invested in each level, there is room to consider the following points:

- As men, we tend to take very seriously the need to secure the base, so that the sustaining of our own existence and security remain firm. Matters such as achievement, esteem, relationships, and other aspects also occupy a prominent place in our agenda. Could it be that, in the midst of the incessant pursuit to satisfy daily needs, the obtaining and accumulation of resources, and the desire to be considered "great," a man may lose his sense of purpose in life?
- Could it be that some are born biologically male, perform the normal functions of a man according to social and cultural assignment, but never develop a clear purpose?
- What things would happen differently if a man could grasp a vision of eternal transcendence that helps him define a sublime purpose in life?

GOD AND THE NEEDS OF MAN

Fortunately, God is not unaware of human needs. In Genesis, chapters 1 and 2, the Holy Scriptures present a very illustrative picture of how precise and accurate God works. It is clearly highlighted that: "So God created man in His own image; in the image of God He created him; male and female He created them" (1:27). In creating man (humanity), He also performed the first wedding: "Then God blessed them, and God said to them: Be fruitful and multiply; fill the earth and subdue it; have dominion over the fish of the sea, over the birds of the air, and over every living thing that moves on the earth."

God is a God of purpose. It is striking to observe all the preparations that God took into account before creating man and establishing the first couple in Eden. Let us look at the process followed:

First day: He created light and separated darkness (1:3–5), giving place to day and night.

Second day: He created the firmament (1:6–8). The beautiful blue sky came onto the scene and was separated from the waters.

Third day: God created the land, the seas, and the vegetation (1:9–13). Here, the Creator commands the waters to gather in one place and the dry land to appear. On that portion called earth, He then created herbs, trees, fruits, and flowers that not only turned what was once chaos into paradise, but also ensured what would later be the source of physical food for what would be His masterpiece.

Fourth day: The great lights—sun, moon, and stars—were placed on that day (1:14–19). How would vegetation be sustained if there were no sun? The God of purpose, owner of the original design of everything that exists, first guarantees the satisfaction of the need and then creates the beneficiary. We know that the sun is a source of light and heat,[6] and without it there is no photosynthesis.[7] It influences the regulation of plant cycles[8] and is essential for human development.[9]

Fifth day: The waters were filled with marine animals, and birds were also created (1:20–23). Fish of every size and color, as well as birds with their beautiful singing, gave life and beauty to these environments.

Sixth day: God created land animals in extraordinary variety. Later, He created man, His masterpiece (1:24–31). His body is composed of 11 main systems that work in a harmonious and impressive way. God's planning and order are unique. They are shown not only in how complex the human body is, but also in everything that was previously made to satisfy his needs and guarantee his survival. And, of course, everything that was made could have been much simpler and still functional. For the enjoyment of man, what God created had absolute perfection and extreme beauty.

When I think about the details of creation, I sometimes ask myself: how much would one night cost in a resort with those characteristics? My life takes on special meaning when I see the way God made all things.

Seventh day: God rested and sanctified that day (2:1–3). After finishing His creative work, God took time to rest—not because He needed it, but to give a lesson of love to His creatures (Mark 2:27).

MAN: YESTERDAY AND TODAY

When man came from the hands of the Creator, what was his moral condition and character like? When contemplating His work, God Himself declared that what He had made was "very good" (Genesis 1:31). Expanding on this, a North American author highlights that "his nature was in harmony with the will of God. His mind was capable of understanding divine things.

His affections were pure, his appetites and passions were under the control of reason... He was holy and felt happy in bearing the image of God and in perfect obedience to the will of the Father."[10]

When sin entered our world (Genesis 3), everything changed. The wickedness of man increased to such a degree that "every intent of the thoughts of his heart was only evil continually" (Genesis 6:5). This was not accidental. Previously, in Genesis 3:15, the Bible points to something that is much more than two lines of thought. The text says: "And I will put enmity between you and the woman, and between your seed and her Seed; He shall bruise your head, and you shall bruise His heel."

The text speaks of the serpent and the woman. Then it speaks of two seeds. In Revelation 12:9 and 20:2, reference is made to the "ancient serpent," named as the devil or Satan. On the other hand, the figure of the woman was not simply a reference to Eve. Figuratively, it represented the people of God (Galatians 4:4; Revelation 12:1, 5), and her seed, in a special way, represented Christ, as presented in Galatians 3:16 and Hebrews 2:14. In the end, the serpent and Christ head two schools of leadership and two curricula of masculinity. In some way, every human being has developed a vision oriented under the shadow of one of these two schools.

Impact of Father Absence: Three Critical Areas

Various studies have pointed out that the absence of the father figure in the home may be associated with multiple consequences in the lives of children. These consequences can be grouped into three main areas: poverty, risk behaviors, and emotional weakness.

1. Poverty and Socioeconomic Vulnerability

The absence of the father in the home is associated with a greater risk of economic difficulties.

- In the United States, one out of every three children (appro-ximately 24 million) lives without his biological father at home.
- Children who grow up in fatherless homes are almost four times more likely to live in poverty.[11]
- Father absence is also related to more fragile family trajectories,[12] as those who grew up without an active father presence show higher probabilities of facing instability in their own marriages in adulthood.[13]

These statistics reflect how the father's presence can influence not only the immediate well-being of the home, but also future family stability.

2. Risk Behaviors

The absence of the father figure has also been linked to a greater likelihood of risk behaviors during adolescence.

- The majority of youth in correctional facilities come from homes where the father figure is absent.[14]
- Adolescent girls who grow up without a father at home are two to three times more likely to experience early pregnancy.[15]
- A positive relationship has been found between father absence and higher consumption of alcohol and drugs.[16]
- Various studies have documented an intergenerational pattern in incarceration. A significant proportion of incarcerated individuals report having had a father who was imprisoned, and parental incarceration is associated with a higher probability of delinquency in the next generation.[17] It is estimated that 5 million children in the United States have had a father incarcerated at some point.[18]

This data suggests that the presence of a stable father figure can act as a protective factor against various social risks.

3. Emotional Weakness and Psychosocial Development

The absence or presence of the father also significantly influences emotional and psychological development.

- Studies have found that father absence is associated with a higher risk of depression and suicidal thoughts in some youth.[19]
- On the contrary, when the father actively participates, children show better academic performance, especially in reading, mathematics, and social skills.[20]
- Children raised with an engaged father demonstrate greater self-control and emotional regulation.[21]

These findings suggest that father involvement contributes to a more solid emotional structure. Beyond material provision, the father figure significantly influences character formation, emotional security, and life purpose in future generations.

From Man to Man

Choose three concrete points to state, in your opinion, what these statistics mean for us, as men today:

1. ______________________________.
1. ______________________________.
2. ______________________________.

Despite what has been previously stated, the man of today struggles to move forward. At times he is perceived as lacking purpose; at other times, he gives the impression of being in the middle of a labyrinth, searching unsuccessfully for the way out.

What does the man of today struggle against? He struggles against cognitive deterioration produced by financial stress. He struggles against addictions and harmful habits. He struggles against the emotional illiteracy that is so prevalent in our century. He struggles in solitude because he does not have a mentor to guide him to the next level. He struggles to look good in the midst of a society that holds high expectations regarding what hegemonic masculinity should fulfill. And now, this man of value, who is defining his purpose in life, finds himself undecided between what he fears and what he desires.

What does man fear? He fears homophobia. He fears being a man, yet not being considered as such. The social curriculum of many societies has determined that whoever qualifies to be considered a man must: renounce expressing emotions, avoid tasks considered feminine, be competitive, strong, tough, successful, powerful, and not allow himself to be taken advantage of by anyone. He must give the impression that he can do everything (even if it is not true), that he knows everything (or invents it), and demonstrate that he is superior to other men in as many things as possible, including the ability to conquer women—who are viewed by many as someone who exists to serve men.

How do you want to see yourself? Between what is feared and what is desired: where do you place yourself? The answer to this question may not be simple.

The Man of Purpose

Having been born with the male sex and truly being a man of purpose are two very different things. However, being able to combine both realities is, in truth, a great privilege. The man of purpose does not live concerned about appearances or about fulfilling the social curriculum. Before acting, he asks himself: What does God think about this?

What God had in mind when creating man can help us in the process of understanding and defining his purpose. Genesis 1:26 tells us:

> "Then God said, 'Let Us make man in Our image, according to Our likeness; and let them have dominion over the fish of the sea, over the birds of the heavens, over the cattle, over all the earth, and over every creeping thing that creeps on the earth.'"

In this text, an important word is "have dominion."Other versions use expressions such as: "let them have dominion," [22] "rule,"[23] "reign,"[24] "govern."[25] It must be highlighted that, when this command was given, God had humanity in mind and not exclusively the male sex. Verse 28 is even more explicit when it says:

> "And God blessed them, and God said to them, 'Be fruitful and multiply; fill the earth and subdue it; and have dominion...'"

Here, in both (Adam and Eve), the additional concept of subduing is emphasized, understood as the mutual responsibility to lead, organize, and administer. Who was to dominate whom? Some use this passage to point out that, since man was created first, he was given the order to "dominate," and that since Eve was the one who sinned and then led man to fall, she would have fallen into an inferior position and therefore should be governed by the husband. It is clear that Genesis 1:27 shows a difference when it says: "male and female." But what kind of difference is this?

This difference does not refer to levels, such as superiority or inferiority. It does not allude to functions, such as leadership or submission. The text shows a difference of biological order, not of hierarchies. In Genesis 1:28, God's blessing rested upon both. This blessing included both in the responsibility of procreation. Likewise, the position of stewardship was shared equally.[26] Furthermore, the biblical text is clear in detailing that the command to dominate and subdue was not given in relation to people, but to things.

As can be observed, before man was created, God made sure to take all necessary care so that he would have what was needed for life. It is interesting to note that God does not create randomly, but rather had a purpose for man and placed him in the world so that he could fulfill it.

What Does the New Testament Say?

Ephesians 5:22–24 brings to the table a passage that has been considered by many as sweet and appealing for the man. The apostle directs his attention to the woman who has chosen well in marriage, indicating to her: "Wives, submit to your own husbands, as to the Lord..." It would be dangerous to read this passage in isolation. When looking at the immediate context and the text itself, we find several very revealing elements:

1. Verse 21 introduces the point by establishing a fair guideline: "Submitting to one another in the fear of God." This indicates that the enrichment of the marital relationship does not occur when one or the other tries to be better, greater, or more important, but when there is mutual submission in marital interaction.

2. In the same way that the woman is instructed to be subject to her husband, verses 25 and 28–29 clarify the role of the husband by saying: "Husbands, love your wives, just as Christ also loved the church and gave Himself for her..." What, then, does loving the woman imply?

   a. The command to love the woman is not subject to the ability to understand her logically. It is founded on a principle that emanates from the heart of Christ: love.

   b. In the style of hegemonic masculinity, male leadership is often applauded when it is authoritarian and coercive. However, the leadership presented in the passage is different, for it follows the model of Christ.

   c. The apostle Paul compares the relationship of husband and wife with the mystical union of Christ with the church. He highlights that Christ loved the church—and what did He do for her? It says that, after exhausting His best efforts without achieving the expected success, He gave Himself and gave His life for her.

I know that you are probably thinking that what Christ did for His wife was something scandalous. I cannot deny that I think so too. Who else would have thought of it? Precisely for that reason, He became the best model for man.

From Man to Man

1. After exhausting all His efforts to attract her and restore her, without obtaining the desired success, what did Christ do for the church?
2. Who is the church/wife of Christ? How did the wife behave with Christ for Him to treat her that way?
3. Who hurt whom more: I to Christ, or my wife to me?
4. What could we say against our wives after contemplating Christ's treatment toward His church?
5. Do the previous points mean that the man should lose his family leadership or headship function, or do they rather clarify it? Justify your answer.

____

____

____

____

Christ is the head of the church, which is His wife. In His condition as Head: "He loves, serves, unifies, nourishes, shares. The idea is not an authoritarian attitude in which power is abused. To be head implies honor and responsibility."[27] With His example, Christ calls the man to wise, balanced, selfless, and loving leadership. He invites the man to value his wife and consider her his companion, his "suitable helper." Therefore, he identifies with her pain, her complaints, her struggles, weaknesses, anguish, and tensions.[28]

In light of such a style, the submission of the woman is not something automatic nor something she should consider as a servile or degrading act, but a duty she exercises through the inalienable use of her own will, in an atmosphere of respect and harmony.

The man was created with the purpose of reflecting the character of God and being His agent or representative in His beautiful creation. God not only created him but also desired to have a close relationship with him, and through that relationship man would find satisfaction, purpose, and fulfillment. Hence, man's supreme purpose is reached when he decides to live for the glory of God.

Isaiah 43:7 affirms it: "All who are called by My name; for My glory I have created them, I have formed them, yes, I have made them." A positive and impactful masculinity is revealed not only when one has great dreams or extraordinary plans, but when one chooses to live for the glory of God.

The Man of Purpose

The man of purpose, by following the model of Christ, is also identified by his leadership of service. Walking with Jesus, His disciples once had an interesting discussion about leadership. Their individual goal was to position themselves advantageously in order to obtain an important place in the kingdom that, according to their expectations, their Master would establish. Upon hearing them, Jesus said:

> "...The kings of the nations exercise lordship over them, and those who exercise authority over them are called benefactors. But you shall not be so. On the contrary, he who is greatest among you, let him be as the younger, and he who governs as he who serves" (Luke 22:24–26).

At that moment, Jesus uses a figure familiar to them and asks: "For who is greater, he who sits at the table, or he who serves? Is it not he who sits at the table? Yet I am among you as the One who serves" (verse 27).

His example was consistent in demonstrating that the purpose is not to be served, but to serve. Such words and modeling were not familiar to them, but that example provided the necessary inspiration that would transform their lives forever.

Where, then, does true masculine greatness lie? It is not hidden in power, success, or the demonstration of dominance. It lies rather in the capacity to love, to serve, and to sacrifice for his wife and for his children.

Purpose and Service

The leadership of Jesus, defined by His purpose and intentional service, was not formed during the days of His pilgrimage on this earth. On the penultimate day of creation, after God gave form to everything that exists, it was time to create man. All the works that preceded him were created for his enjoyment and delight, as if a loving father were to receive his son after a long vacation and say: "Son, look at all the things I have prepared for you."

However, the case here is much more complex. Until that moment, everything had been created by the express command of the Creator's voice. Now, the Maker of the universe bends down to the earth and, with His own hands, without haste, gives form to His masterpiece.

On one hand, let us look at the Creator. Who is He? He is the King of the universe, eternal and all-knowing. Nothing takes Him by surprise, for He lives in an eternal present. His knowledge is a preview of history, and history is the confirmation of His wisdom, without being modified by it.

On the other hand, let us observe the creature. Before God intervened, he was simply dust, but the marvelous work of God gave him form, life, meaning, and purpose. God could have created an automaton; instead, He made a being with free will, capable of making his own decisions.

Now the matter becomes even more complex: the King of the universe humbles Himself to the dust to create a being with free will, who can decide whatever he pleases—even to deny his Maker and live with his back turned to Him. It draws my attention that, even so, God decided to create man, despite knowing that this man would one day kill Him.

The Holy Scriptures say: "But when the fullness of the time had come, God sent forth His Son, born of a woman..." (Galatians 4:4).

Isaiah 53 describes the Son of God as the Suffering Servant, saying: "Despised and rejected by men, a Man of sorrows and acquainted with grief... like a lamb that is led to the slaughter... and as a sheep before its shearers is silent, so He opened not His mouth."

The Son of God came to planet Earth, which He had created with such care and beauty, and here He was humiliated, beaten, wounded, judged, and condemned outside the terms of the law. Finally, He was sentenced to death (John 3:16).

The million-dollar question is: who killed Him? That same man whom He had created long ago in Eden, despite knowing that sooner or later the time would come when he would kill Him.

You may ask yourself: if He knew it, then why did He create him? God created him because He not only has love; He is love. I do not know a single human being who would have done what God did. We can be more than grateful that God works with a model in which human rules do not apply.

Likewise, in Philippians 2:5–11, the apostle Paul adds another point worthy of our careful attention:

> "Let this mind be in you which was also in Christ Jesus, who, being in the form of God, did not consider it robbery to be equal with God, but made Himself of no reputation, taking the form of a servant, and coming in the likeness of men. And being found in appearance as a man, He humbled Himself and became obedient to the point of

> death, even the death of the cross. Therefore God also has highly exalted Him... that at the name of Jesus every knee should bow..."

What Jesus did among men and for men distinguishes Him above all. The day will come when every knee will bow in respect, worship, and reverence before the One who did everything for us.

Based on this text, another question arises: when did Jesus become great among men? Was it when He was above in heaven, or when He descended to earth to save humanity? The Bible indicates that it was not when He was above, but when He came down and became man in order to save man.

Why is this model so important? Because it contrasts with the popular tendency of our days. Everywhere it is common to see people seek honor more than responsibility, the higher position before humility. However, before exaltation comes humiliation, and before honor, responsibility makes its appearance.

Sometimes what man receives is far from what he expects or deserves. There is no doubt about that. However, if God was willing to love in the way He did, the man who desires to reflect His character is intentional in learning to love, serve, and forgive, in spite of...

A New Masculine

The model provided by Christ, when contrasted with the style of masculinity that in many cases has shaped us or that we are accustomed to seeing, highlights that today a new masculine is needed. One in which a fit brain looks better and is more sought after than a fit body; where quantity loses the throne to quality, and where presence is worth more than appearance.

I must be honest in saying that not only a new masculine is needed. A new feminine is also needed. There is no doubt that many women have not only accepted as normal the pattern they complain about, but often promote it. When a man does not meet the expectations of masculinity that popularly define him, there is no lack of a woman who turns away and, with her voice or gesture, suggests: "No... that is not a man."

On the other hand, who is the one who raises the man? The man is raised by a woman, whose worldview needs to be transformed, because before her work can produce a new result, she must have new ideas that help generate new family patterns, inspired by the infallible model of Jesus Christ. Then society will have a new product: a man of value, emotionally fit, whose purpose will be to live a life that gives glory to his Model.

In the century of light, men are still sought in the shadows—men who have trained their brains to be fit and whose lives follow the model of the Creator. In a time of so vast technological advances, there is no software that can program the mind, but there is a higher power that can not only inspire but also transform the heart and bring it closer in likeness to the character of the One who formed it in the beginning.

It is through a constant relationship with his Creator that man can be renewed until he becomes emotionally fit. Then the brightness of those qualities that distinguish him among many around him will become more evident. This will serve as a mold for the formation of new generations and for the construction of a better society.

Let Us Consult the Bible and Complete the Idea

1. "For the Son of Man did not come to be ___________, but to ______________, and to give His life as a _____________ for many" (Mark 10:45).
2. "For I have given you an ___________, that you should do as I have done to you" (John 13:15).

3. "Husbands, ______________ your wives and do not be ______________ with them" (Colossians 3:19).
4. "Husbands, likewise, dwell with them with _________________, giving honor to the ______________ ________________" (1 Peter 3:7).
5. "Let this mind be in ______________ which was also in ______________ ____________________" (Philippians 2:5).

How to Train the Brain to Be a Fit Man on the Inside

To be an integrally fit man, according to the Creator's model, is an integral process that lasts a lifetime. It is not an impossible mission, but a conscious and transformative journey. What are its main characteristics?

1. He is clear about who the supreme authority is

This man was able to understand that Christ is the supreme authority and has given Him the opportunity to sit and rule on the throne of his heart. In this way, he gave himself the opportunity to know who Jesus is: His life, His lordship, His mission, and the role that every human being plays in His plan. He chose to draw near to the source of infinite wisdom and to allow divine omniscience to direct his life.

It is when the man comes to know Christ that he discovers who he truly is. Then, what he most longs for is to allow Him to occupy the throne of his heart.

> What happens when Jesus takes control?
>
> Where there is chaos, He brings order.
>
> Where there is darkness, He makes His light shine.

At times, there are strongholds that have served as protection and also as an excuse to preserve a poor and deficient character, marked by trauma, pain, and tragedy. Those strongholds succumb before the presence of Christ, being transformed into an Eden of blessing and an extraordinary power for good.

It is only necessary to begin a relationship of faith with Him, which will grow as it is cultivated. The Bible says: "But as many as received Him, to them He gave the right to become children of God, to those who believe in His name" (John 1:12).

The word right is synonymous with power, and it conveys the idea of a dynamic, transformative force. That power will be on your side to help you move forward. To have Him on your side is to have everything, because against Him nothing and no one can prevail.

2. He exercises his privileged function as spiritual guide of the family

The man represents values and principles. He is the one in charge of bringing Christ to his family and leading his family into the presence of God.

Just as in the people of Israel a sacrifice was offered every morning and every evening, each day it is the man's privilege to inspire faith and model it among his own. It is God's plan that this principle extends from generation to generation (Deuteronomy 4:4–9).

3. He recognizes and demonstrates appreciation for the woman

Whether single or married, his life is linked to a lady: mother, sister, wife, or daughter.

He shows appreciation for her effort, recognizes her value, and understands how much everything she does for the well-being of the home means. A daily dose of affirmation can tear down a ton of bitterness.

4. He is a servant leader

He is not satisfied only when he is served, but also when he serves.

He is intentional in helping at home. He knows that the man who serves more is not less of a man, nor does the one who leads by example lose influence.

5. He takes time to speak, but more to listen

Every couple needs time to converse. Not about things or about other people, but about themselves.

It is estimated that 15–20 minutes daily of meaningful communication can make a great difference in the marital relationship. The lady needs to express herself and be listened to attentively. That regular exercise strengthens her emotional health and the well-being of the bond.

6. He is a present husband

Physical presence is important, but not sufficient.

The emotionally fit man understands the importance of giving moral support to his wife, assuring her that she can count on him. He values her dreams, that degree she could not complete, that life project she visualized with excitement, and he seeks to stamp upon them his decided support.

When the years pass and she looks back, he wants her to be able to see a husband who walked by her side, who inspired her to go through difficulties and to move on to the next level.

7. He establishes the correct order in his relationships

I have seen husbands who were excellent fathers, but not so good as partners. As a consequence, the relationship wears down and degenerates into an emotional divorce.

For that reason, the man of value prioritizes the husband-wife relationship, and together they work to give the best to their children.

8. He values intimacy as a unique experience

He does not take for granted that things will simply happen. He acts with intention and wisdom, preparing the emotional and spiritual atmosphere for that special moment. He understands that intimacy is not routine, but encounter.

9. He trusts and is trustworthy

Trust is one of the fundamental pillars of every meaningful relationship.

The man of value knows how to trust, but even more, he is worthy of trust. Trust does not arise automatically; it is cultivated with the seeds of daily behavior.

The one who has earned trust rarely has to ask for it. He loves his wife and, if he had to compare her with any other woman, his own always comes out winning. He sees himself by her side in youth and in old age, in good times and in bad. His plans are with her and for her.

10. He knows who his wife is and prepares for the end of the movie

He not only knows who his wife is, but what she is: a princess.

She is daughter of the King of Kings. For that reason, he conducts himself in a relationship of mutual respect and affection. He is conscious that very soon both of them will have an encounter with the King, and he desires to be able to present her before Him on that expected occasion.

The Integrally Fit Man

The integrally fit man discovered who he is and found the purpose that gave meaning to his life. He learned to regulate his emotions and does not depend on the domination of his environment for his self-realization.

He knows how to speak and he knows how to listen. He articulates his ideas with wisdom. For him, sex is more than a physical union; it is also an encounter where emotions, intellect, and spirit come together.

When it comes to money, he manages it with responsibility. He organizes himself, plans his future, and teaches his family to manage it prudently.

Although the challenges of life and work commitments may keep him temporarily distant, he preserves with his partner and his family a unique, appreciable, and growing emotional connection.

He became fit because he trained his brain for a masculinity with purpose. He was not afraid to begin or complete this journey, and he was helped by the vast experience of the best Trainer: Christ Jesus, the supreme model of redeemed and restored masculinity.

Now I encourage you to keep training. Make it a life habit. And by your example, allow its multiplying effect to bring health to many relationships.

Endnotes:

1. Maslow, A. H. (1943). *A theory of human motivation. Psychological Review,* 50(4), 370–396. https://psychclassics.yorku.ca/Maslow/motivation.htm

2. Skinner, B. F. (1938). *The behavior of organisms: An experimental analysis.* Appleton-Century.

3. Skinner, B. F. (1953). *Science and human behavior. Macmillan.*

4. Freud, S. (1953). *The interpretation of dreams. Standard Edition,* Vols. 4–5. Hogarth Press.

5. Freud, S. (1961). *The ego and the id.* Standard Edition, Vol. 19.

6. *Encyclopaedia Britannica.* (n.d.). Photosynthesis. In Encyclopaedia Britannica. Retrieved January 23, 2026, from https://www.britannica.com/science/photosynthesis

7. NASA Earthdata. (n.d.). *Photosynthetically active radiation.* NASA Earthdata. Retrieved January 23, 2026, from https://www.earthdata.nasa.gov/topics/biosphere/photosynthetically-active-radiation

8. U.S. Department of Agriculture, Agricultural Research Service. (n.d.). *Major chapter 4* [PDF]. https://www.ars.usda.gov/ARSUserFiles/30980500/Major%20Chapter%204.pdf

9. Blume C, Garbazza C, Spitschan M. Effects of light on human circadian rhythms, sleep and mood. Somnologie (Berl). 2019 Sep;23(3):147-156. doi: 10.1007/s11818-019-00215-x. Epub 2019 Aug 20. PMID: 31534436; PMCID: PMC6751071.

10. White, E. G. (2012). *Patriarcas y profetas.* Asociación Casa Editora Sudamericana.

11. National Fatherhood Initiative. (2021, November 10). *The father absence crisis [Infographic].* Fatherhood.org. https://www.fatherhood.org/championing-fatherhood/the-father-absence-crisis-infographic?.com

12. McLanahan S, Tach L, Schneider D. *The Causal Effects of Father Absence. Annu Rev Sociol.* 2013 Jul;39:399-427. doi: 10.1146/annurev-soc-071312-145704. PMID: 24489431; PMCID: PMC3904543.

13. Teachman, J. (2003). *Childhood living arrangements and marital dissolution. Journal of Marriage and Family*, 65(2), 354–367.

14. Office of Juvenile Justice and Delinquency Prevention. (2018). *Selected findings from the 2018 National Juvenile Justice Data Analysis* (OJJDP Juvenile Residential Facility Census Databook – Selected Findings). U.S. Department of Justice. https://ojjdp.ojp.gov/publications/jrfc-2018-selected-findings.pd

15. Ellis BJ, Bates JE, Dodge KA, Fergusson DM, Horwood LJ, Pettit GS, Woodward L. Does father absence place daughters at special risk for early sexual activity and teenage pregnancy? Child Dev. 2003 May-Jun;74(3):801-21. doi: 10.1111/1467-8624.00569. PMID: 12795391; PMCID: PMC2764264.

16. Hemovich, V., Lac, A., & Crano, W. (2011). *Understanding early-onset drug and alcohol outcomes. Journal of Community Psychology,* 39(7), 791–805.

17. Dobbie, W., Grönqvist, H., Niknami, S., Palme, M., & Priks, M. (2018). *The intergenerational effects of parental incarceration* (NBER Working Paper No. 24186). National Bureau of Economic Research. https://doi.org/10.3386/w24186

18. Murphey, D., & Cooper, P. M. (2015). *Parents behind bars: What happens to their children?* Child Trends. https://www.researchgate.net/publication/285234895_Parents_Behind_Bars_What_Happens_to_Their_Children

19. Langhinrichsen-Rohling, J., et al. (2002). Risk for suicidal ideation in youth. *Suicide and Life-Threatening Behavior,* 32(4), 343–357.

20. Lamb, M. E. (2010). *The Role of the Father in Child Development* (5th ed.). Wiley.

21. Rollè L, Gullotta G, Trombetta T, Curti L, Gerino E, Brustia P, Caldarera AM. Father Involvement and Cognitive Development in Early and Middle Childhood: A Systematic Review. Front Psychol. 2019 Oct 25;10:2405. doi: 10.3389/fpsyg.2019.02405. PMID: 31708843; PMCID: PMC6823210.

22. English Standard Version (ESV).

23. New International Version (NIV)

24. New Living Translation (NLT)

25. New English Bible (NEB)

26. Davidson, R. M. (2007). *Flame of Yahweh: Sexuality in the Old Testament.* Peabody, MA: Hendrickson Publishers.

27. Taylor, G. D., & Taylor, W. D. (1984). *La familia auténticamente cristiana.* Portavoz Evangélico.

28. Getz, G. A. (1980). *La medida de una familia.* CLIE.

Emocionalmente Fit

Final Evaluation

Diagnosis of Progress and Transformation

Name: ______________________________

Date: ______________________________

Initial score (if applicable): ________________

Instructions:

Respond considering your reality after the process lived.

Score	Meaning
1	Does not reflect my reality
2	Minimal change
3	Partial change
4	Consistent change
5	Solid and evident change

I. Self-perception and identity

Item	1	2	3	4	5
I have greater clarity about who I am as a man.					
My identity is less dependent on external approval.					
I recognize my strengths and weaknesses honestly.					
I am more willing to ask for help when I need it.					

My decisions are more aligned with principles.					
I have progressed in defining my personal purpose.					

Subtotal: ____ / 30

II. Emotional management

Item	1	2	3	4	5
I better identify what I feel.					
I express emotions with greater control.					
I manage anger in a healthier way.					
I regulate my emotions better under pressure.					
have reduced resentments or impulsive reactions.					
I demonstrate greater self-control.					

Subtotal: ____ / 30

III. Relationship with the mother

Item	1	2	3	4	5
I have progressed in emotionally healing this relationship.					
I better understand its impact on my life.					
I have worked on forgiveness (if necessary).					

My relational patterns have improved.					
My treatment toward women has evolved positively.					
I can process this relationship with maturity.					

Subtotal: ____ / 30

IV. Relationship with the girlfriend or wife

Item	1	2	3	4	5
My girlfriend / wife feels more secure with me.					
I have improved my ability to listen.					
I express affection and respect with greater consistency.					
I handle conflicts better.					
I have corrected harmful patterns.					
My way of loving is more mature and responsible.					

Subtotal: ____ / 30

V. Fatherhood (sons and daughters)

Item	1	2	3	4	5
I am more intentional in my role as a father.					
My presence transmits greater emotional security.					

I have improved communication with my children.					
I correct with greater balance.					
I am forming character, not only behavior.					
I am a better example of a man for them.					

Subtotal: ____ / 30

VI. Communication

Item	1	2	3	4	5
I listen better than before.					
I express disagreements with respect.					
I communicate with greater clarity.					
I have learned to ask for forgiveness.					
I handle difficult conversations with maturity.					
My communication generates less conflict.					

Subtotal: ____ / 30

VII. Sexuality and integrity

Item	1	2	3	4	5
I have greater control over my impulses.					

I have established healthier boundaries.					
My way of seeing women has changed positively.					
I am more honest with myself.					
My decisions reflect my values.					
I have progressed in personal integrity.					

Subtotal: ____ / 30

VIII. Money and responsibility

Item	1	2	3	4	5
I manage my financial decisions better.					
I am less impulsive with money.					
I am more financially responsible.					
My decisions in this area reflect vision.					
I have improved financial habits.					
I am more disciplined in this area.					

Subtotal: ____ / 30

IX. Character and purpose

Item	1	2	3	4	5
I live with greater intention.					
I am more disciplined.					
I keep my word better.					

I take responsibility without excuses.					
I have greater clarity of purpose.					
I am committed to growing.					

Subtotal: ____ / 30

GLOBAL SCORE

Final total: ________ / 270

PROGRESS CALCULATION

Initial score: ________

Final score: ________

Progress: ________ points

Interpretation Scale (out of 270 points)

Score Level	Level
243 – 270	Emotionally fit man
216 – 242	High development
189 – 215	Good progress
162 – 188	Vulnerable area
Less than 162	Needs intervention

EVIDENCE OF TRANSFORMATION

Mark what you have evidenced:

- ☐ I changed a specific emotional reaction
- ☐ I restored a relationship

- ☐ I asked for forgiveness
- ☐ I established boundaries
- ☐ I changed habits
- ☐ I improved discipline
- ☐ I made different decisions

FINAL REFLECTION

Before I was:

____

____

____

Today I am:

____

____

I need to continue working on:

____

____

____

MY DECLARATION OF HONOR:

"Today I not only understand more... but, by the grace of God, today I am different. My goal is to continue growing and to become each day more like my model par excellence: Christ Jesus."

Other Works by the Author

1. New Clothes for Your Mind – Regular Edition

This book is helping many people reconnect with themselves, uncover and work through important wounds from their past, and find the strength needed to move forward. In its pages, you will learn how to improve your attitude, manage anger, and relate more wisely when dealing with difficult people and their offenses. It also provides valuable resources to face anxiety, stress, depression, jealousy, and forgiveness. It is a key tool for emotional growth. Available at www.hopenestfamily.com and www.amazon.com

2. New Clothes for Your Mind – School Edition, Teacher's Guide

After reading the regular edition, many parents and educators suggested developing a special classroom version. This edition provides teachers with a didactic guide that includes multiple activities, turning the teaching of these topics into a dynamic, participatory, and enriching experience. Available at www.amazon.com

3. New Clothes for Your Mind - School Edition, Student Workbook

This resource accompanies students in discovering their inner world, helping them better understand their emotions, relate in a healthy way with others, and manage conflicts wisely. It offers practical tools for a balanced life, enabling them to face stress, jealousy, anxiety, depression, loneliness, and forgiveness with greater wisdom. Schools, families, and communities will greatly benefit from this material designed specifically for the educational context. Available at www.amazon.com

4. Spark Up Your Marriage

At a time when divorce and marital crises are on the rise, this work helps couples understand the dynamics of married life and offers practical tools to address conflicts that prevent the relationship from functioning fully. An inspiring resource to strengthen relationships and renew married life. Available at www.amazon.com

5. Proactive Leadership

If you are a leader or church member who feels comfortable and is not interested in seeing the community come to your congregation and become engaged, this book may not be for you. But if your desire is to generate a true spiritual revival in your church and help the community discover the value and appeal of connecting with God and it, then this book deserves to be read, studied, and discussed by your leadership team. Available at www.amazon.com

6. Dear Young Lovers... – 2nd Edition

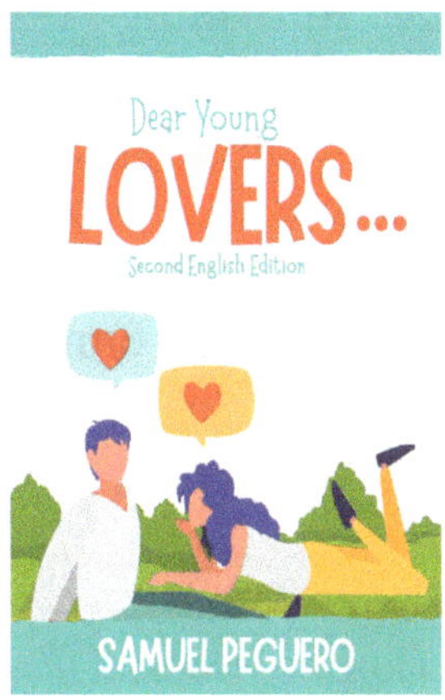

No one gets married with the intention of failing; however, the desire to succeed in love is not enough. This book offers not only advice but also a clear vision for making wise decisions. By addressing life-shaping topics, it promotes critical thinking and guides young people toward building strong families. It is an ideal resource for youth, singles, parents who want to guide their children, teachers, psychologists, and youth leaders. Available at www.hopenestfamily.com and www.amazon.com

7. Dear Young Lovers... – Instructor's Guide

This guide is a valuable resource for teachers, counselors, parents, and youth leaders working with adolescents, young adults, and singles. It develops themes such as friendship, dating, risks associated with this stage of life, and preparation for marriage, connecting learning with real-life application. It can be used alongside the regular version of the book (5th edition in Spanish or 2nd edition in English). Available at www.hopenestfamily.com and www.amazon.com

8. Families with Hope (also available in English, French, and Portuguese)

A refreshing resource that combines teachings on family life with biblical principles for healthy living. It presents a compilation of twelve lessons designed to be studied, reflected upon, and shared within the family, strengthening bonds and nurturing hope at home. Available at www.hopenestfamily.com

Coming Soon...

Cocktail of Emotions: An Exuberant Dance of Emotions in Stories from the Old Testament

Books in Spanish / Libros en Espanol:

1. Ropa Nueva para Tu Mente - versión regular.

Este libro está ayudando a muchas personas a encontrarse consigo mismas, descubrir y trabajar heridas importantes de su historia y encontrar la fortaleza necesaria para avanzar. En sus páginas aprenderás cómo mejorar tu actitud, manejar la ira y relacionarte con mayor sabiduría al tratar con personas difíciles y sus ofensas. Además, ofrece valiosos recursos para enfrentar la ansiedad, el estrés, la depresión, los celos y el perdón. Es una herramienta clave para el crecimiento emocional. Disponible en www.hopenestfamily.com y www.amazon.com

2. Ropa Nueva para Tu Mente - versión escolar, edición del profesor.

Después de leer la versión regular, varios padres y educadores sugirieron desarrollar una edición especial para el aula. Esta versión proporciona al docente una guía didáctica con múltiples actividades que convierten la enseñanza de estos temas en una experiencia dinámica, participativa y enriquecedora. Disponible en www.amazon.com

3. Ropa Nueva para Tu Mente - versión escolar, edición para el estudiante.

Este recurso acompaña al estudiante en el descubrimiento de su mundo interior, ayudándole a comprender mejor sus emociones, relacionarse de forma saludable con los demás y manejar con prudencia los conflictos. Ofrece herramientas prácticas para una vida equilibrada, permitiendo enfrentar con mayor sabiduría el estrés, los celos, la ansiedad, la depresión, la soledad y el perdón. Escuelas, familias y comunidades se verán grandemente beneficiadas con este material preparado especialmente para el contexto educativo. Disponible en www.amazon.com

4. Ponle Chispa a Tu Matrimonio.

En un tiempo en que los divorcios y las crisis matrimoniales van en aumento, esta obra ayuda a comprender la dinámica de la vida de pareja y ofrece herramientas prácticas para enfrentar los conflictos que impiden que la relación funcione plenamente. Un recurso inspirador para fortalecer la relación y renovar la vida matrimonial. Disponible en www.hopenestfamily.com y www.amazon.com

5. Caldo de Pollo para la Familia.

Una colección de relatos inspiradores que promueven la dicha y la unión familiar. De lectura sencilla, cautivante y profundamente aleccionadora. Disponible en www.hopenestfamily.com y www.amazon.com

6. Poemas y Reflexiones para Alimentar el Alma.

Los poemas y reflexiones de esta colección nacen del corazón del autor y buscan transmitir un mensaje de esperanza, fe y renovación espiritual. Disponible en www.hopenestfamily.com y www.amazon.com

7. Liderazgo Proactivo.

Si usted es un líder o miembro de iglesia que se siente cómodo y no está interesado en que la comunidad llegue a su congregación y se entusiasme con ella, tal vez este libro no sea para usted. Pero si su deseo es generar una verdadera revolución espiritual en su iglesia y ayudar a que la comunidad descubra el valor y el atractivo de acercarse a ella, entonces este libro merece ser leído, estudiado y discutido por su junta directiva. Disponible en librerías IADPA en América Latina y en línea.

8. Charlas para Jóvenes Enamorados – 5.ª edición.

Nadie se casa con el propósito de fracasar; sin embargo, el deseo de triunfar en el amor no es suficiente. Este libro no solo ofrece consejos, sino también una visión clara para tomar decisiones sabias. Al abordar temas formadores de destino, promueve el pensamiento crítico y orienta a los jóvenes hacia la construcción de familias sólidas. Es un recurso ideal para jóvenes, solteros, padres que desean orientar a sus hijos, profesores, psicólogos y líderes juveniles. Disponible en www.hopenestfamily.com y www.amazon.com

9. Charlas para Jóvenes Enamorados – Guía para el Instructor.

Esta guía constituye un recurso valioso para docentes, orientadores, padres y líderes juveniles que trabajan con adolescentes, jóvenes y solteros. Desarrolla los temas de amistad, noviazgo, riesgos asociados a esta etapa de la vida y preparación para el matrimonio, conectando el aprendizaje con la vida real. Puede utilizarse junto a la versión regular del libro (5.ª edición en español o 2.ª edición en inglés). Disponible en www.hopenestfamily.com y www.amazon.com

10. Familias con Esperanza (disponible también en inglés, francés y portugués).

Un material refrescante que combina enseñanzas sobre vida familiar con principios bíblicos del buen vivir. Presenta un compendio de doce lecciones diseñadas para estudiar, reflexionar y compartir en familia, fortaleciendo los vínculos y la esperanza en el hogar. Disponible en ... Disponible en www.hopenestfamily.com

11. Emocionalmente Fit

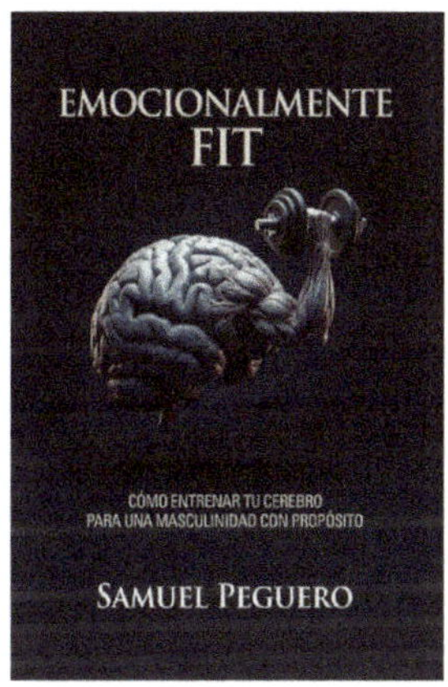

Muchos varones se esfuerzan por desarrollar su cuerpo, su carrera y su capacidad de proveer, pero pocas veces han sido entrenados para comprender y gobernar su mundo emocional. Emocionalmente Fit nace precisamente para llenar ese vacío. Esta obra integra principios de biología, neurociencia, sociología, psicología y teología para ofrecer una explicación científica y práctica del cerebro masculino. Aprenderás a trabajar tus puntos débiles, potenciar tus fortalezas y reprogramar aquellas áreas que han limitado tu crecimiento. Disponible en www.hopenestfamily.com www.libros.market (Sudamérica y España) y www.amazon.com

Muy pronto....

Cóctel de Emociones: Exuberante Danza de Emociones en Historias del Antiguo Testamento.

About the Author

Dr. Samuel Peguero is a recognized specialist in family life, educator, speaker, and author, whose work for more than 20 years has impacted thousands of people in diverse cultural contexts. He holds a solid academic background that includes a bachelor's degree in Theology, master's degrees and specializations in Family Counseling, Marital Therapy, as well as Clinical Pastoral Education and Psychotherapy. He also holds a doctorate in Family Life from Andrews University, in Michigan, United States. He currently serves as Director of the Department of Family and Singles Ministries in the Greater New York Conference. Additionally, he is an adjunct professor at Andrews University and at Universidad Adventista Dominicana.

As an author, he has developed an extensive editorial production that includes ten works, widely used in different educational, ministerial, and therapeutic settings. Among his most outstanding titles are Ropa Nueva para tu Mente, in its various versions (regular and school), Charlas para Jóvenes Enamorados (also published in English as Dear Young Lovers), Liderazgo Proactivo, and Ponle Chispa a tu Matrimonio, among others. His materials are characterized by a practical, formative, and transformative approach, integrating contributions from psychology, neuroscience, and spirituality. Through his publications, he has contributed significantly to the development of training programs aimed at youth, couples, men, parents, teachers, and leaders.

Throughout his career, Dr. Peguero has worked directly with families from more than 130 cultures, developing a deep understanding of relational dynamics in multicultural contexts. His experience includes leading seminars, retreats, certification programs, and mentoring spaces in different countries. In addition, he has had a notable presence in media—radio, print, and television—where he has shared practical guidance on family, relationships, and emotional health. His voice has been established as a reliable reference for those seeking to restore, strengthen, and build their homes on solid principles. He has the blessing of being the husband of Wirdis and, together with her, the proud father of two extraordinary daughters: Lisbeth and Angie.

www.ingramcontent.com/pod-product-compliance
Lightning Source LLC
Chambersburg PA
CBHW050526240926
51846CB00003B/712